计算机技术
开发与应用丛书

Unity游戏单位驱动开发

张寿昆 著

清华大学出版社
北京

内 容 简 介

本书是一本介绍 Unity 游戏开发技术的实用教程，旨在帮助读者掌握游戏开发中常用的技术和方法。全书共分为 7 章，内容涵盖了 Unity 的输入系统、数学基础、相机控制、物理检测、动画系统、寻路算法及游戏单位的驱动方法等多方面。

本书既适合初学者入门，也对有多年工作经验的开发者具有参考价值。通过阅读本书，读者将能够掌握 Unity 游戏开发的核心技术和方法，为创作高质量的游戏作品奠定坚实的基础。

版权所有，侵权必究。举报：010-62782989，beiqinquan@tup.tsinghua.edu.cn。

图书在版编目(CIP)数据

Unity 游戏单位驱动开发 / 张寿昆著. -- 北京：清华大学出版社，2025.2.
(计算机技术开发与应用丛书). --ISBN 978-7-302-67959-2

Ⅰ. TP317.6

中国国家版本馆 CIP 数据核字第 2025TP2450 号

责任编辑：赵佳霓
封面设计：吴 刚
责任校对：刘惠林
责任印制：丛怀宇

出版发行：清华大学出版社
    网　　址：https://www.tup.com.cn,https://www.wqxuetang.com
    地　　址：北京清华大学学研大厦 A 座　　邮　　编：100084
    社 总 机：010-83470000　　邮　　购：010-62786544
    投稿与读者服务：010-62776969，c-service@tup.tsinghua.edu.cn
    质量反馈：010-62772015，zhiliang@tup.tsinghua.edu.cn
    课件下载：https://www.tup.com.cn,010-83470236
印 装 者：北京瑞禾彩色印刷有限公司
经　　销：全国新华书店
开　　本：186mm×240mm　　印　　张：15.5　　字　　数：350 千字
版　　次：2025 年 3 月第 1 版　　印　　次：2025 年 3 月第 1 次印刷
印　　数：1～1500
定　　价：69.00 元

产品编号：106004-01

前言
PREFACE

随着游戏产业的蓬勃发展，Unity作为一款强大的跨平台游戏引擎，受到了越来越多游戏开发者的青睐。Unity以其灵活性和易用性为开发者提供了实现创意想法的广阔舞台。

本书旨在为广大Unity游戏开发者提供一本全面深入的实用教程，帮助大家更好地掌握游戏开发中的核心技术。通过本书的学习，读者将能够更加熟练地运用Unity引擎进行游戏开发，打造出更加精彩、有趣的游戏作品。无论是初学者还是有一定基础的读者都能从本书中获得宝贵的经验和启示。

本书在写作过程中使用的Unity版本为2020.3.16f1c1，因为不同版本的API可能会略有不同，因此建议读者在学习过程中使用相同的版本，扫描目录上方的二维码可下载本书源码。

本书具体章节安排如下。

第1章详细介绍了Unity的输入系统，包括旧输入系统和新输入系统的使用方法，帮助读者快速掌握游戏开发中常见的输入处理方式。

第2章介绍了Unity开发中的数学基础，包括Mathf数学运算工具类、向量和矩阵的概念及基本运算，为后续的游戏开发提供了必要的数学基础。

第3章讲解了多种类型相机控制组件的实现，包括第一人称视角、第三人称视角和自由视角等多种相机控制方式，并介绍了Unity中强大的Cinemachine系统，帮助读者轻松实现复杂的相机控制。

第4章介绍了Unity中的物理检测，通过应用实例介绍射线投射检测、球体投射检测、盒体重叠检测等物理检测的使用方法，并详细介绍了如何借助Gizmos实现物理检测的可视化，让读者能够更加直观地理解物理检测工作的原理。

第5章深入讲解了Unity的动画系统，包括动画剪辑、动画状态机、动画事件、动画曲线、BlendShape和反向动力学等多个方面，帮助读者轻松实现游戏中的复杂动画效果。

第6章介绍了几种常见的自动寻路实现方式，包括Unity内置的Navigation、A星寻路和流场寻路等算法，帮助读者掌握游戏开发中常用的路径规划和导航技术。

第7章详细介绍了游戏单位的驱动方法，包括用户人物角色的驱动、敌方战斗单位的驱动、载具驱动等，为读者提供了开发工作中常见的游戏单位控制技术的实现方法和技巧。

在写作过程中，作者得到了家人和朋友的帮助，在此表示感谢。同时，感谢清华大学出版社赵佳霓编辑的细心指导。

由于作者知识水平有限，书中难免存在疏漏之处，欢迎读者批评指正。

最后，真诚地希望本书能够成为您学习 Unity 游戏开发的良师益友，并祝愿您在游戏开发的道路上越走越远，创造出更多精彩的游戏作品。

张寿昆

2024 年 12 月

目 录
CONTENTS

本书源码

第 1 章　输入系统 ·· 1

1.1　旧输入系统 Input Manager ·· 2
　　1.1.1　获取鼠标按键输入 ·· 2
　　1.1.2　 获取物理按键输入 ·· 3
　　1.1.3　获取虚拟轴输入 ·· 7
　　1.1.4　获取按钮的输入 ·· 9
　　1.1.5　获取触摸屏的触摸输入 ··· 10
　　1.1.6　Input Manager 配置 ·· 12
　　1.1.7　XBox 手柄设备的输入 ·· 14

1.2　新输入系统 Input System ·· 27
　　1.2.1　基于旧输入系统做兼容 ··· 27
　　1.2.2　Input Action Asset 配置文件 ··· 37
　　1.2.3　Player Input 组件 ··· 40

第 2 章　数学基础 ··· 43

2.1　Mathf ·· 43
　　2.1.1　常量 ··· 43
　　2.1.2　三角函数 ·· 44
　　2.1.3　插值函数 ·· 44
　　2.1.4　最值与限制函数 ··· 45
　　2.1.5　幂、平方根、对数函数 ··· 46

2.2　向量 ·· 47
　　2.2.1　向量加减 ·· 47
　　2.2.2　向量数乘 ·· 48

    2.2.3 向量插值 ·· 48
    2.2.4 向量点乘与叉乘 ·· 48
  2.3 矩阵 ··· 51
    2.3.1 矩阵的基本运算 ·· 52
    2.3.2 变换矩阵 ·· 53

第 3 章 相机控制 ·· 57
  3.1 第一人称类型相机 ·· 57
  3.2 第三人称类型相机 ·· 59
    3.2.1 通过角色朝向控制视角 ···································· 59
    3.2.2 通过用户输入控制视角 ···································· 62
  3.3 自由控制类型相机 ·· 66
    3.3.1 观察者视角控制 ·· 66
    3.3.2 漫游视角控制 ·· 71
  3.4 Cinemachine ·· 73
    3.4.1 基于虚拟相机实现第三人称视角 ···························· 73
    3.4.2 轨道路径与推轨相机 ······································ 75
    3.4.3 在 Timeline 中控制镜头 ·································· 78

第 4 章 物理检测 ·· 85
  4.1 射线投射检测 ··· 85
    4.1.1 获取鼠标单击地面位置 ···································· 86
    4.1.2 游戏物体事件响应系统 ···································· 90
  4.2 球体投射检测 ··· 95
  4.3 盒体重叠检测 ··· 98
  4.4 物理检测可视化 ··· 103
    4.4.1 盒体重叠检测可视化 ···································· 104
    4.4.2 盒体投射检测可视化 ···································· 106
    4.4.3 球体投射检测可视化 ···································· 108

第 5 章 动画系统 ·· 111
  5.1 动画剪辑 ··· 111
    5.1.1 录制关键帧 ·· 112
    5.1.2 创建和编辑关键帧 ······································ 112
    5.1.3 外部导入的动画资产 ···································· 112
  5.2 动画状态机 ··· 114

- 5.2.1 Animator 窗口 ·· 114
- 5.2.2 动画状态 ··· 115
- 5.2.3 动画过渡 ··· 116
- 5.2.4 混合树 ·· 118
- 5.3 动画事件 ·· 120
  - 5.3.1 Animation Clip Event ·· 121
  - 5.3.2 State Machine Behaviour ···································· 121
- 5.4 动画曲线 ·· 122
- 5.5 BlendShape ·· 123
- 5.6 反向动力学 ··· 124
  - 5.6.1 Animator IK ·· 125
  - 5.6.2 人物角色脚部放置方案 ······································ 126
  - 5.6.3 IK 权重值曲线烘焙 ·· 131

第 6 章 寻路算法 ·· **140**

- 6.1 Navigation ··· 140
  - 6.1.1 导航网格 ··· 140
  - 6.1.2 导航网格代理 ·· 142
  - 6.1.3 导航网格障碍物 ··· 145
  - 6.1.4 网格外链接 ··· 146
- 6.2 A 星寻路 ·· 147
  - 6.2.1 地图数据 ··· 147
  - 6.2.2 计价方式 ··· 150
  - 6.2.3 邻节点搜索方式 ··· 151
  - 6.2.4 算法实现 ··· 153
  - 6.2.5 寻路组件 ··· 156
  - 6.2.6 寻路代理 ··· 157
  - 6.2.7 路径优化 ··· 160
  - 6.2.8 地图编辑器 ··· 162
- 6.3 流场寻路 ·· 166
  - 6.3.1 流场 ·· 167
  - 6.3.2 算法实现 ··· 168
  - 6.3.3 寻路组件 ··· 170
  - 6.3.4 寻路代理 ··· 173
- 6.4 八叉树寻路 ··· 174

第 7 章 游戏单位驱动 … 184

7.1 用户人物角色驱动 … 184
7.1.1 基于刚体组件实现人物角色驱动 … 184
7.1.2 基于角色控制器组件实现人物角色驱动 … 188

7.2 人物角色行为 … 194
7.2.1 跳跃 … 194
7.2.2 滑行 … 196
7.2.3 翻越 … 201
7.2.4 掩体行为 … 205

7.3 敌方战斗单位驱动 … 213
7.3.1 有限状态机 … 213
7.3.2 敌方战斗单位 AI … 221

7.4 汽车驱动 … 225
7.4.1 车轮碰撞器 … 225
7.4.2 驱动类型 … 227
7.4.3 车辆转向 … 230
7.4.4 行驶速度 … 230
7.4.5 ABS 与 ASR … 231
7.4.6 尾气排放 … 232
7.4.7 车辆音效 … 235
7.4.8 撞击变形 … 235

第 1 章　输　入　系　统

Unity 在 2019 版本中推出了新输入系统 Input System 的预览版,于 2020.1 版本中改为正式版,从此便有了新旧两套输入系统。

尽管有了新的输入系统,在创建一个新的 Unity 项目时,默认使用的仍然是旧输入系统。如果想要使用新输入系统,则需要在 Project Settings 窗口中的 Player 页中找到 Active Input Handling 设置选项进行切换,如图 1-1 所示。Input Manager 选项对应旧输入系统,Input System Package 选项对应新输入系统,Both 选项则表示同时使用新旧两套输入系统。

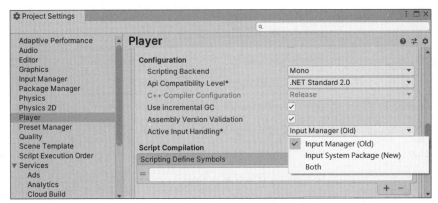

图 1-1　Active Input Handling

切换 Active Input Handling 选项时,如果使用的输入系统发生变更,则会弹出一个确认弹窗,提示需要重启 Unity 编辑器,以便使变更生效,如图 1-2 所示。

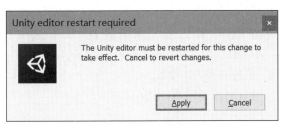

图 1-2　提示弹窗

1.1 旧输入系统 Input Manager

在使用旧输入系统时，获取用户的输入需要使用 Input 类，该类是一个专门用于处理玩家输入操作的类，它能够接收和处理来自键盘、鼠标、触摸屏和游戏摇杆等输入设备的信号。Input 类中包含的静态方法及对应的作用见表 1-1。

表 1-1　Input 类中的静态方法及作用

静态方法	作用
GetAccelerationEvent()	获取上一帧期间进行的特定加速度测量(不分配临时变量)
GetAxis()	获取指定的虚拟轴的值
GetAxisRaw()	获取指定的虚拟轴的值(未应用平滑过渡)
GetButton()	获取指定的虚拟按钮是否被持续按下
GetButtonDown()	获取指定的虚拟按钮是否开始被按下
GetButtonUp()	获取指定的虚拟按钮是否被抬起
GetJoystickNames()	获取在 Input Manager 中配置的与摇杆索引对应的输入设备名称的列表
GetKey()	获取指定按键是否被持续按下
GetKeyDown()	获取指定按键是否开始被按下
GetKeyUp()	获取指定按键是否被抬起
GetMouseButton()	获取指定的鼠标按键是否被持续按下
GetMouseButtonDown()	获取指定的鼠标按键是否开始被按下
GetMouseButtonUp()	获取指定的鼠标按键是否被抬起
GetTouch()	获取屏幕设备触摸输入信息
IsJoystickPreconfigured()	是否已预先配置某个特定的游戏摇杆模型(Linux)
ResetInputAxes()	重置所有输入(所有虚拟轴和按钮都恢复为 0，持续一帧时长)

1.1.1 获取鼠标按键输入

鼠标按键分为鼠标左键、鼠标右键及滚轮按键，获取这些按键输入的方法包括 GetMouseButtonDown()、GetMouseButton() 和 GetMouseButtonUp()，分别表示按键开始被按下、按键持续被按下、按键被抬起，参数 button 为 int 类型，表示按键值，0 表示鼠标左键，1 表示鼠标右键，2 表示鼠标滚轮按键，代码如下：

```
public static bool GetMouseButtonDown (int button);
public static bool GetMouseButton (int button);
```

```
public static bool GetMouseButtonUp (int button);
```

在按键开始被按下的那一帧，GetMouseButtonDown()方法的返回值为true，在按键被抬起的那一帧，GetMouseButtonUp()方法的返回值为true，在按键被抬起之前，GetMouseButton()方法的返回值始终为true。

以获取鼠标左键的输入为示例，代码如下：

```
//第1章/InputManagerExample.cs

using UnityEngine;

public class InputManagerExample : MonoBehaviour
{
    private void Update()
    {
        GetMouseButtonInputExample();
    }
    private void GetMouseButtonInputExample()
    {
        if (Input.GetMouseButtonDown(0))
            Debug.Log("鼠标左键开始被按下");
        if (Input.GetMouseButton(0))
            Debug.Log("鼠标左键持续被按下");
        if (Input.GetMouseButtonUp(0))
            Debug.Log("鼠标左键被抬起");
    }
}
```

将该示例组件挂载于场景中的某个游戏物体上，运行程序，按下鼠标左键并抬起，查看控制台，结果如图1-3所示。

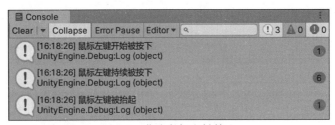

图1-3　获取鼠标左键输入

1.1.2　获取物理按键输入

KeyCode是一个枚举，里面定义了各物理按键对应的枚举值，这里的物理按键既包括键盘上的按键,也包括遥控器、手柄等设备上的按键,详情见表 1-2。

表 1-2　KeyCode 枚举值详解

枚 举 值	详　　解
None	未指定按键
Backspace	Backspace 按键
Delete	删除按
Tab	Tab 按键
Clear	清除键
Return	Enter 键
Pause	暂停键
Escape	退出键
Space	空格键
Keypad0…Keypad9	数字小键盘 0～9 按键
KeypadPeriod	数字小键盘"."按键
KeypadDivide	数字小键盘"/"按键
KeypadMultiply	数字小键盘"*"按键
KeypadMinus	数字小键盘"－"按键
KeypadPlus	数字小键盘"＋"按键
KeypadEnter	数字小键盘 Enter 按键
KeypadEquals	数字小键盘"＝"按键
UpArrow、DownArrow、RightArrow、LeftArrow	向上、向下、向右、向左方向按键
Insert	插入键
Home	主页键
End	结束键
PageUp	向上翻页按键
PageDown	向下翻页按键
F1…F15	F1～F15 功能键
Alpha0…Alpha9	字母数字键盘上的 0～9 按键
Exclaim	叹号键"!"
DoubleQuote	双引号键
Hash	井号键"#"

续表

枚 举 值	详 解
Dollar	美元符号键"$"
Percent	百分比符号键"%"
Ampersand	与符号键"&"
Quote	单引号键
LeftParen	左圆括号键"("
RightParen	右圆括号键")"
Asterisk	星号键"*"
Plus	加号键"+"
Comma	逗号键","
Minus	减号键"-"
Period	句点键"."
Slash	斜杠键"/"
Colon	冒号键":"
Semicolon	分号键";"
Less	小于号键"<"
Equals	等号键"="
Greater	大于号键">"
Question	问号键"?"
At	"@"键
LeftBracket	左方括号键"["
RightBracket	右方括号键"]"
Backslash	反斜杠键"\"
Caret	光标键"^"
Underscore	下画线键"_"
BackQuote	反引号键"`"
A...Z	字母 A~Z 按键
LeftCurlyBracket	左大括号键"{"
RightCurlyBracket	右大括号键"}"
Pipe	竖线键"|"

续表

枚 举 值	详 解
Tilde	波浪符号键"～"
Numlock	数字锁定键
CapsLock	大写锁定键
ScrollLock	滚动锁定键
LeftShift	左 Shift 键
RightShift	右 Shift 键
LeftControl	左 Ctrl 键
RightControl	右 Ctrl 键
LeftAlt	左 Alt 键
RightAlt	右 Alt 键
LeftCommand	左 Command 键（macOS）
RightCommand	右 Command 键（macOS）
LeftApple	左 Command 键（macOS）
RightApple	右 Command 键（macOS）
LeftWindows	左 Windows 键
RightWindows	右 Windows 键
AltGr	Alt Gr 键
Help	帮助键
Print	打印键
SysReq	Sys Req 键
Break	Break 键
Menu	菜单键
Mouse0…7	鼠标左、右、中及其他附加按钮
JoystickButton0…19	任何游戏摇杆上的 0～19 按钮
Joystick1…8Button0…19	第 1～8 个游戏摇杆上的 0～19 按钮

获取这些物理按键输入的方法包括 GetKeyDown()、GetKey() 和 GetKeyUp()，分别表示按键开始被按下、按键持续被按下、按键被抬起，参数 key 表示按键的枚举值，name 表示按键的名称，为了使代码易于维护，通常使用枚举值表示按键，代码如下：

```
public static bool GetKeyDown (string name);
```

```
public static bool GetKeyDown (KeyCode key);
public static bool GetKey (string name);
public static bool GetKey (KeyCode key);
public static bool GetKeyUp (string name);
public static bool GetKeyUp (KeyCode key);
```

显然,这些方法和获取鼠标按键输入的方法的使用方式一致。以获取键盘上字母 A 键的输入为示例,代码如下:

```
//第 1 章/InputManagerExample.cs

private void Update()
{
    GetKeyCodeInputExample();
}
private void GetKeyCodeInputExample()
{
    if (Input.GetKeyDown(KeyCode.A))
        Debug.Log("键盘 A 键开始被按下");
    if (Input.GetKey(KeyCode.A))
        Debug.Log("键盘 A 键持续被按下");
    if (Input.GetKeyUp(KeyCode.A))
        Debug.Log("键盘 A 键被抬起");
}
```

运行程序,按下键盘上的字母 A 键并抬起,结果如图 1-4 所示。

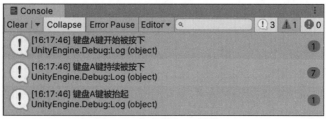

图 1-4　获取键盘 A 键输入

1.1.3　获取虚拟轴输入

在 Input Manager 的默认设置中包含的虚拟轴有 Horizontal、Vertical、Mouse X、Mouse Y 及 Mouse ScrollWheel 等,如图 1-5 所示。

Horizontal、Vertical 分别用于获取用户在水平、垂直方向上的输入值。当按下键盘上的 W 键或上方向键时,Vertical 虚拟轴将返回一个正值,当按下键盘上的 S 键或下方向键

图 1-5 Input Manager 设置

时，Vertical 虚拟轴将返回一个负值。对应地，当按下键盘上的 D 键或右方向键时，Horizontal 虚拟轴将返回一个正值，当按下键盘上的 A 键或左方向键时，Horizontal 虚拟轴将返回一个负值。

Mouse X、Mouse Y 虚拟轴分别用于获取用户的鼠标在水平、垂直方向上的输入值。当用户的鼠标向右移动时，Mouse X 虚拟轴将返回一个正值，当用户的鼠标向左移动时，Mouse X 虚拟轴将返回一个负值。对应地，当用户的鼠标向上移动时，Mouse Y 虚拟轴将返回一个正值，当用户的鼠标向下移动时，Mouse Y 虚拟轴将返回一个负值。

Mouse ScrollWheel 虚拟轴则用于获取用户的鼠标滚轮的滚动值。当鼠标滚轮向上滚动时，Mouse ScrollWheel 虚拟轴将返回一个正值，当鼠标滚轮向下滚动时，Mouse ScrollWheel 虚拟轴将返回一个负值。

获取这些虚拟轴的输入有 GetAxis() 和 GetAxisRaw() 两种方法，参数 axisName 表示虚拟轴的名称，代码如下：

```
public static float GetAxis (string axisName);
public static float GetAxisRaw (string axisName);
```

这两种方法的区别在于，GetAxis() 方法返回的值应用了平滑过渡，而 GetAxisRaw() 方法返回的值未应用平滑过渡。

以 Horizontal 虚拟轴为例，如果使用 GetAxis() 方法，则当用户按下键盘 D 键或右方向键时方法的返回值将从 0 逐渐过渡到 1，松开按键后，方法的返回值逐渐恢复为 0，而如果使用 GetAxisRaw() 方法，则当用户按下键盘 D 键或右方向键时方法的返回值始终为 1，松开按键后，方法的返回值直接恢复为 0。

示例代码如下：

```csharp
//第 1 章/InputManagerExample.cs
private void Update()
{
    GetAxisInputExample();
}
private void GetAxisInputExample()
{
    float horizontal=Input.GetAxis("Horizontal");
    float vertical=Input.GetAxisRaw("Vertical");
    Debug.Log(string.Format("[{0},{1}]", horizontal, vertical));
}
```

1.1.4　获取按钮的输入

Input 类中用于获取按钮输入的方法包括 GetButtonDown()、GetButton()、GetButtonUp()，分别表示按钮开始被按下、按钮持续被按下、按钮被抬起，参数 buttonName 表示按钮的名称，代码如下：

```csharp
public static bool GetButtonDown (string buttonName);
public static bool GetButton (string buttonName);
public static bool GetButtonUp (string buttonName);
```

在 Input Manager 的默认设置中有一个名为 Jump 的按钮，该按钮对应的键盘按键为空格键，如图 1-6 所示，以获取该按钮的输入为示例，代码如下：

```csharp
//第 1 章/InputManagerExample.cs
private void Update()
{
    GetButtonInputExample();
}
private void GetButtonInputExample()
{
    if (Input.GetButtonDown("Jump"))
        Debug.Log("Jump 按钮开始被按下");
    if (Input.GetButton("Jump"))
        Debug.Log("Jump 按钮持续被按下");
    if (Input.GetButtonUp("Jump"))
        Debug.Log("Jump 按钮被抬起");
}
```

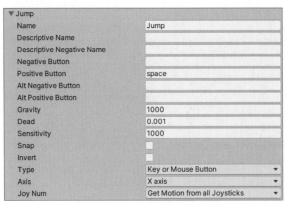

图 1-6　Jump 按钮配置

运行程序,按下键盘上的空格键并抬起,获取 Jump 按钮输入的结果如图 1-7 所示。

图 1-7　获取 Jump 按钮输入

1.1.5　获取触摸屏的触摸输入

Touch 是一个结构体,用于描述用户在触摸屏设备上的触摸输入信息,触摸输入通过 Input 类中的静态方法 GetTouch()获取,代码如下:

```
public static Touch GetTouch (int index);
```

触摸输入可能会有多个,参数 index 作为索引,用于获取指定的触摸输入。触摸输入的数量通过 Input 类中的静态变量 touchCount 获取。当 touchCount 为 1 时,表示单点触摸,当 touchCount 大于或等于 2 时,表示多点触摸。

通过 Touch 结构体中的 position 可以获得触摸点的位置,通过 phase 可以获得触摸的阶段,它是一个枚举值,详情见表 1-3,示例代码如下:

```
//第 1 章/InputManagerExample.cs

private void Update()
{
    GetTouchInputExample();
}
```

```csharp
private void GetTouchInputExample()
{
    if (Input.touchCount ==1)
    {
        Touch touch=Input.GetTouch(0);
        Debug.Log(string.Format(
            "单点触摸 触摸坐标{0}", touch.position));
    }
    else if (Input.touchCount>=2)
    {
        for (int i=0; i<Input.touchCount; i++)
        {
            Touch touch=Input.GetTouch(i);
            Debug.Log(string.Format(
                "多点触摸 [{0}] {1}", i, touch.phase));
        }
    }
}
```

表 1-3　TouchPhase 详解

枚 举 值	详　解
Began	手指开始触摸屏幕
Moved	手指在触摸屏上移动
Stationary	手指正在触摸屏幕,但尚未移动
Ended	手指从触摸屏上抬起(触摸的最终阶段)
Canceled	取消了对触摸的跟踪

当调试触摸屏设备上的输入时,需要构建对应的包体并部署到设备中进行测试,而构建包体通常需要消耗大量的时间,通过 Unity Remote 应用程序则可以简化这种测试工作,省去进行完整构建的麻烦。

Unity Remote 通过将应用程序与 Unity Editor 进行连接,可以实时获取设备上的输入,Editor 中的可视输出也将被发送到设备屏幕上。

iOS 端可以在 App Store 中下载 Unity Remote,而 Android 端可以在 Google Play 中进行下载。下载并安装完成后运行应用程序,如图 1-8 所示,通过 USB 数据线将设备连接到计算机。需要注意的是,移动端设备需要打开开发者模式和 USB 调试模式。

在 Unity 中打开 Project Settings 窗口,找到 Unity Remote 的相关设置,以 Android 端为例,将 Device 选项更改为 Any Android Device,如图 1-9 所示。

在 Build Settings 中 Platform 也需要设置为对应的平台,如图 1-10 所示。

图 1-8　Unity Remote 应用程序

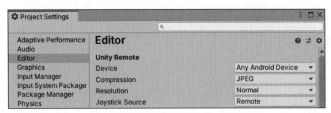

图 1-9　Unity Remote Settings

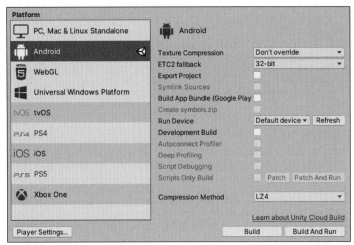

图 1-10　Build Settings Platform

所有设置完成后运行程序，可以发现已经可以通过移动端设备获取触摸输入了，如图 1-11 所示。

1.1.6　Input Manager 配置

在 Input Manager 中可以为项目自定义输入轴及其关联操作，每个轴包含的属性及对应功能见表 1-4。

如果要将按键映射到轴，则需要在 Positive Button 或 Negative Button 中输入按键对应的名称，按键遵循的命名约定见表 1-5。

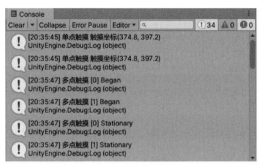

图 1-11　获取触摸输入

表 1-4　轴的属性及功能

属　　性	功　　能
Name	轴的名称
Descriptive Name、Descriptive Negative Name	这些值已经弃用，不起作用
Negative Button、Positive Button	分别用于沿负向、正向推动轴的控件
Alt Negative Button、Alt Positive Button	分别用于沿负向、正向推动轴的备用控件
Gravity	当不存在输入时，轴下降到中性点的速度
Dead	在应用程序对移动操作进行记录之前，用户需要移动摇杆的距离。在运行时，所有设备在该范围内的输入将被视为 null
Sensitivity	轴向目标值移动的速度
Snap	如果启用该属性，则当按下对应于反方向的按钮时轴值将被重置为 0
Invert	是否反转
Type	轴的类型，包含 3 种，分别为 Key or Mouse Button、Mouse Movement、Joystick Axis
Axis	用于控制该轴的连接设备的轴
JoyNum	用于控制该轴的连接游戏摇杆

表 1-5　按键遵循的命名约定

按　　键	命　名　约　定
字母键	a,b,c…
数字键	1,2,3…
箭头键	up,down,left,right
小键盘按键	[1],[2],[3],[+],[equals]…

续表

按　键	命 名 约 定
修饰键	right shift、left shift、right ctrl、left ctrl、right alt、left alt、right cmd、left cmd
特殊键	backspace、tab、return、escape、space、delete、enter、insert、home、end、page up、page down
功能键	f1、f2、f3…

鼠标左键、右键、滚轮按键的命名分别为 mouse 0、mouse 1、mouse 2。任何游戏摇杆上特定按钮的命名为 joystick button 0、joystick button 1 等。如果是特定游戏摇杆上的特定按钮，则需要命名为 joystick 1 button 0、joystick 1 button 1、joystick 2 button 0，以此类推。

1.1.7　XBox 手柄设备的输入

在 1.1.2 节和 1.1.3 节中分别介绍了如何获取物理按键和虚拟轴的输入，本节以 XBox 手柄设备为示例，介绍如何获取手柄设备的输入。

当 XBox 手柄设备与 PC 连接、断开连接或重新连接时，可以在控制台看到连接状态日志，如图 1-12 所示。

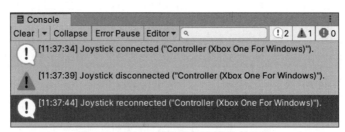

图 1-12　XBox 手柄设备连接状态日志

设备接入后，可以通过 Input 类中的静态方法 GetJoystickNames() 查看在 Input Manager 中配置的与摇杆索引对应的输入设备名称列表，代码如下：

```
private void Start()
{
    string[] joystickNames=Input.GetJoystickNames();
    for (int i=0; i<joystickNames.Length; i++)
    {
        Debug.Log(joystickNames[i]);
    }
}
```

运行结果如图 1-13 所示。

XBox 手柄设备的按键说明如图 1-14 所示，首先要建立它们与 Unity 中物理按键或虚拟轴的对应关系，详情见表 1-6。

图 1-13 查看设备名称

图 1-14 XBox 手柄设备的按键说明

表 1-6 XBox 手柄设备按键与 Unity 中物理按键或虚拟轴的对应关系表

按键/摇杆	KeyCode/Axis	具 体 值
视图按钮	KeyCode	JoystickButton6
菜单按钮	KeyCode	JoystickButton7
左扳机键	Axis	LT
右扳机键	Axis	RT
左缓冲键	KeyCode	JoystickButton4
右缓冲键	KeyCode	JoystickButton5
左摇杆	Axis	LeftStickHorizontal、LeftStickVertical
左摇杆按钮	KeyCode	JoystickButton8
X 按钮	KeyCode	JoystickButton2
Y 按钮	KeyCode	JoystickButton3
A 按钮	KeyCode	JoystickButton0
B 按钮	KeyCode	JoystickButton1
右摇杆	Axis	RightStickHorizontal、RightStickVertical
右摇杆按钮	KeyCode	JoystickButton9
方向盘	Axis	DPadHorizontal、DPadVertical

在 1.1.6 节中介绍了如何自定义配置，以 XBox 手柄设备上的 A 按钮为例，根据表 1-6 在 Input Manager 中进行配置，结果如图 1-15 所示。

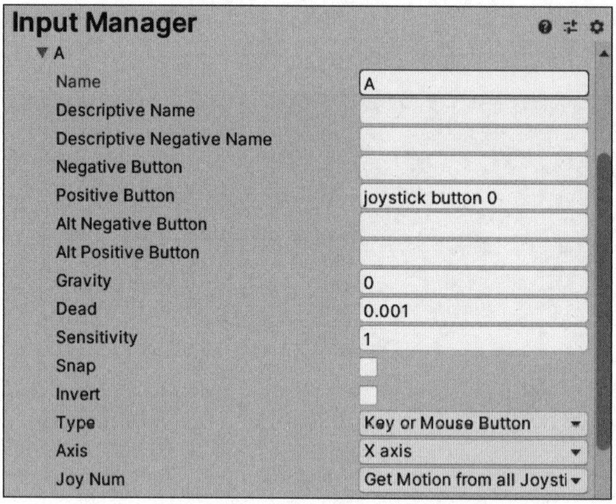

图 1-15　XBox 手柄设备按钮 A 的配置

将手柄上所有的按键配置完成后，在项目根目录中打开 ProjectSettings 文件夹，找到其中的 InputManager.asset 文件，打开该文件，内容如下：

```
%YAML 1.1
%TAG !u! tag:unity3d.com,2011:
--- !u!13 &1
InputManager:
  m_ObjectHideFlags: 0
  serializedVersion: 2
  m_Axes:
  - serializedVersion: 3
    m_Name: LeftStickHorizontal
    descriptiveName:
    descriptiveNegativeName:
    negativeButton:
    positiveButton:
    altNegativeButton:
    altPositiveButton:
    gravity: 0
    dead: 0.19
    sensitivity: 1
    snap: 0
    invert: 0
    type: 2
```

```
    axis: 0
    joyNum: 0
  - serializedVersion: 3
    m_Name: LeftStickVertical
    descriptiveName:
    descriptiveNegativeName:
    negativeButton:
    positiveButton:
    altNegativeButton:
    altPositiveButton:
    gravity: 0
    dead: 0.19
    sensitivity: 1
    snap: 0
    invert: 1
    type: 2
    axis: 1
    joyNum: 0
  - serializedVersion: 3
    m_Name: A
    descriptiveName:
    descriptiveNegativeName:
    negativeButton:
    positiveButton:
    altNegativeButton:
    altPositiveButton:
    gravity: 0
    dead: 0.001
    sensitivity: 1
    snap: 0
    invert: 0
    type: 0
    axis: 0
    joyNum: 0
  - serializedVersion: 3
    m_Name: B
    descriptiveName:
    descriptiveNegativeName:
    negativeButton:
    positiveButton: joystick button 1
    altNegativeButton:
```

```
    altPositiveButton:
    gravity: 0
    dead: 0.001
    sensitivity: 1
    snap: 0
    invert: 0
    type: 0
    axis: 0
    joyNum: 0
  - serializedVersion: 3
    m_Name: X
    descriptiveName:
    descriptiveNegativeName:
    neqativeButton:
    positiveButton: joystick button 2
    altNegativeButton:
    altPositiveButton:
    gravity: 0
    dead: 0.001
    sensitivity: 1
    snap: 0
    invert: 0
    type: 0
    axis: 0
    joyNum: 0
  - serializedVersion: 3
    m_Name: Y
    descriptiveName:
    descriptiveNegativeName:
    negativeButton:
    positiveButton: joystick button 3
    altNegativeButton:
    altPositiveButton:
    gravity: 0
    dead: 0.001
    sensitivity: 1
    snap: 0
    invert: 0
    type: 0
    axis: 0
    joyNum: 0
```

```
- serializedVersion: 3
  m_Name: LB
  descriptiveName:
  descriptiveNegativeName:
  negativeButton:
  positiveButton: joystick button 4
  altNegativeButton:
  altPositiveButton:
  gravity: 0
  dead: 0.001
  sensitivity: 1
  snap: 0
  invert: 0
  type: 0
  axis: 0
  joyNum: 0
- serializedVersion: 3
  m_Name: RB
  descriptiveName:
  descriptiveNegativeName:
  negativeButton:
  positiveButton: joystick button 5
  altNegativeButton:
  altPositiveButton:
  gravity: 0
  dead: 0.001
  sensitivity: 1
  snap: 0
  invert: 0
  type: 0
  axis: 0
  joyNum: 0
- serializedVersion: 3
  m_Name: LeftStick
  descriptiveName:
  descriptiveNegativeName:
  negativeButton:
  positiveButton: joystick button 8
  altNegativeButton:
  altPositiveButton:
  gravity: 0
```

```
      dead: 0.001
      sensitivity: 1
      snap: 0
      invert: 0
      type: 0
      axis: 0
      joyNum: 0
  - serializedVersion: 3
      m_Name: RightStick
      descriptiveName:
      descriptiveNegativeName:
      negativeButton:
      positiveButton: joystick button 9
      altNegativeButton:
      altPositiveButton:
      gravity: 0
      dead: 0.001
      sensitivity: 1
      snap: 0
      invert: 0
      type: 0
      axis: 0
      joyNum: 0
  - serializedVersion: 3
      m_Name: View
      descriptiveName:
      descriptiveNegativeName:
      negativeButton:
      positiveButton: joystick button 6
      altNegativeButton:
      altPositiveButton:
      gravity: 0
      dead: 0.001
      sensitivity: 1
      snap: 0
      invert: 0
      type: 0
      axis: 0
      joyNum: 0
  - serializedVersion: 3
      m_Name: Menu
```

```
    descriptiveName:
    descriptiveNegativeName:
    negativeButton:
    positiveButton: joystick button 7
    altNegativeButton:
    altPositiveButton:
    gravity: 0
    dead: 0.001
    sensitivity: 1
    snap: 0
    invert: 0
    type: 0
    axis: 0
    joyNum: 0
  - serializedVersion: 3
    m_Name: LT
    descriptiveName:
    descriptiveNegativeName:
    negativeButton:
    positiveButton:
    altNegativeButton:
    altPositiveButton:
    gravity: 0
    dead: 0.19
    sensitivity: 1
    snap: 0
    invert: 0
    type: 2
    axis: 8
    joyNum: 0
  - serializedVersion: 3
    m_Name: RT
    descriptiveName:
    descriptiveNegativeName:
    negativeButton:
    positiveButton:
    altNegativeButton:
    altPositiveButton:
    gravity: 0
    dead: 0.19
    sensitivity: 1
```

```
      snap: 0
      invert: 0
      type: 2
      axis: 9
      joyNum: 0
    - serializedVersion: 3
      m_Name: DPadHorizontal
      descriptiveName:
      descriptiveNegativeName:
      negativeButton:
      positiveButton:
      altNegativeButton:
      altPositiveButton:
      gravity: 0
      dead: 0.19
      sensitivity: 1
      snap: 0
      invert: 0
      type: 2
      axis: 5
      joyNum: 0
    - serializedVersion: 3
      m_Name: DPadVertical
      descriptiveName:
      descriptiveNegativeName:
      negativeButton:
      positiveButton:
      altNegativeButton:
      altPositiveButton:
      gravity: 0
      dead: 0.19
      sensitivity: 1
      snap: 0
      invert: 1
      type: 2
      axis: 6
      joyNum: 0
    - serializedVersion: 3
      m_Name: RightStickHorizontal
      descriptiveName:
      descriptiveNegativeName:
```

```
    negativeButton:
    positiveButton:
    altNegativeButton:
    altPositiveButton:
    gravity: 0
    dead: 0.19
    sensitivity: 1
    snap: 0
    invert: 1
    type: 2
    axis: 3
    joyNum: 0
  - serializedVersion: 3
    m_Name: RightStickVertical
    descriptiveName:
    descriptiveNegativeName:
    negativeButton:
    positiveButton:
    altNegativeButton:
    altPositiveButton:
    gravity: 0
    dead: 0.19
    sensitivity: 1
    snap: 0
    invert: 1
    type: 2
    axis: 4
    joyNum: 0
```

使用以上内容覆盖 InputManager.asset 文件中默认的内容也可以完成配置。为了便于调用相应方法获取设备输入时传参,需要封装对应关系类,代码如下:

```
//第 1 章/XBox.cs

using UnityEngine;

//<summary>
//XBox 手柄设备按键
//</summary>
public class XBox
{
```

```csharp
//<summary>
//左侧摇杆水平轴
//X axis
//</summary>
public const string LeftStickHorizontal="LeftStickHorizontal";
//<summary>
//左侧摇杆垂直轴
//Y axis
//</summary>
public const string LeftStickVertical="LeftStickVertical";
//<summary>
//右侧摇杆水平轴
//4th axis
//</summary>
public const string RightStickHorizontal="RightStickHorizontal";
//<summary>
//右侧摇杆垂直轴
//5th axis
//</summary>
public const string RightStickVertical="RightStickVertical";
//<summary>
//十字方向盘水平轴
//6th axis
//</summary>
public const string DPadHorizontal="DPadHorizontal";
//<summary>
//十字方向盘垂直轴
//7th axis
//</summary>
public const string DPadVertical="DPadVertical";
//<summary>
//LT
//9th axis
//</summary>
public const string LT="LT";
//<summary>
//RT
//10th axis
//</summary>
public const string RT="RT";
//<summary>
```

```csharp
//左侧摇杆按键
//joystick button 8
//</summary>
public const KeyCode LeftStick=KeyCode.JoystickButton8;
//<summary>
//右侧摇杆按键
//joystick button 9
//</summary>
public const KeyCode RightStick=KeyCode.JoystickButton9;
//<summary>
//A 键
//joystick button 0
//</summary>
public const KeyCode A=KeyCode.JoystickButton0;
//<summary>
//B 键
//joystick button 1
//</summary>
public const KeyCode B=KeyCode.JoystickButton1;
//<summary>
//X 键
//joystick button 2
//</summary>
public const KeyCode X=KeyCode.JoystickButton2;
//<summary>
//Y 键
//joystick button 3
//</summary>
public const KeyCode Y=KeyCode.JoystickButton3;
//<summary>
//LB 键
//joystick button 4
//</summary>
public const KeyCode LB=KeyCode.JoystickButton4;
//<summary>
//RB 键
//joystick button 5
//</summary>
public const KeyCode RB=KeyCode.JoystickButton5;
//<summary>
//View 视图键
```

```csharp
        //joystick button 6
        //</summary>
        public const KeyCode View=KeyCode.JoystickButton6;
        //<summary>
        //Menu 菜单键
        //joystick button 7
        //</summary>
        public const KeyCode Menu=KeyCode.JoystickButton7;
}
```

有了上述对应关系类后,便可以清晰地获取手柄上指定按键的输入,示例代码如下:

```csharp
//第1章/XBoxInputExample.cs

using UnityEngine;

public class XBoxInputExample : MonoBehaviour
{
    private void OnGUI()
    {
        GUILayout.Label(string.Format(" A:{0}", Input.GetKey(XBox.A)));
        GUILayout.Label(string.Format(" B:{0}", Input.GetKey(XBox.B)));
        GUILayout.Label(string.Format(" X:{0}", Input.GetKey(XBox.X)));
        GUILayout.Label(string.Format(" Y:{0}", Input.GetKey(XBox.Y)));
        GUILayout.Label(string.Format(" LB:{0}", Input.GetKey(XBox.LB)));
        GUILayout.Label(string.Format(" RB:{0}", Input.GetKey(XBox.RB)));
        GUILayout.Label(string.Format(" Left Stick Button:{0}",
            Input.GetKey(XBox.LeftStick)));
        GUILayout.Label(string.Format(" Right Stick Button:{0}",
            Input.GetKey(XBox.RightStick)));
        GUILayout.Label(string.Format(" View:{0}",
            Input.GetKey(XBox.View)));
        GUILayout.Label(string.Format(" Menu:{0}",
            Input.GetKey(XBox.Menu)));

        GUILayout.Label(string.Format(" Left Stick:{0}", new Vector2(
            Input.GetAxis(XBox.LeftStickHorizontal),
            Input.GetAxis(XBox.LeftStickVertical))));
        GUILayout.Label(string.Format(" Right Stick:{0}", new Vector2(
            Input.GetAxis(XBox.RightStickHorizontal),
            Input.GetAxis(XBox.RightStickVertical))));
```

```
        GUILayout.Label(string.Format(" D Pad:{0}", new Vector2(
            Input.GetAxis(XBox.DPadHorizontal),
            Input.GetAxis(XBox.DPadVertical))));
        GUILayout.Label(string.Format(" LT:{0}", Input.GetAxis(XBox.LT)));
        GUILayout.Label(string.Format(" RT:{0}", Input.GetAxis(XBox.RT)));
    }
}
```

运行程序,按下手柄上的 X 按钮,结果如图 1-16(a)所示,向上推动右摇杆,结果如图 1-16(b)所示。

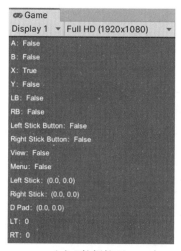
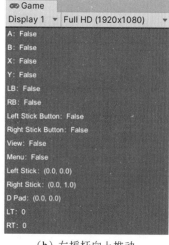

（a）X按钮按下　　　　　　　　（b）右摇杆向上推动

图 1-16　XBox 手柄设备输入测试结果

1.2　新输入系统 Input System

如果要使用新输入系统,则除了需要在设置窗口中对 Active Input Handling 选项进行设置外,还需要在 Package Manager 中找到 Input System 的包体进行导入,如图 1-17 所示。

在导入新输入系统的包体时,如果 Active Input Handling 仍然是旧输入系统的设置选项,编辑器则会弹出告警弹窗,如图 1-18 所示,单击 Yes 按钮后会重启编辑器并切换使用新输入系统。

1.2.1　基于旧输入系统做兼容

新输入系统可以像旧输入系统一样使用,在通过 using 关键字引入 UnityEngine.InputSystem 命名空间后,便可以调用新输入系统中的相关类和方法。

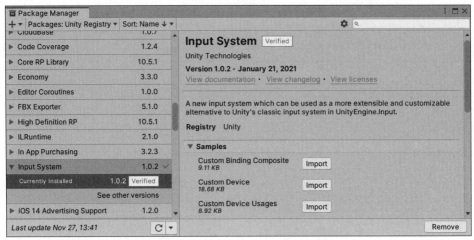

图 1-17　Input System Package

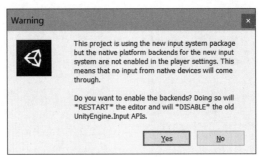

图 1-18　告警弹窗

1. 获取键盘按键输入

Keyboard 类是 InputDevice 类的派生类，表示键盘类型输入设备，该类中包含各键盘按键对应的按键控制类成员属性，这些属性为 ButtonControl 类型或者为 ButtonControl 的派生类型。

ButtonControl 中的 wasPressedThisFrame 属性表示该按键是否在当前帧中开始被按下，isPressed 属性表示该按键是否持续被按下，wasReleasedThisFrame 属性则表示该按键是否在当前帧中被抬起。

以获取键盘上的 A 键的输入为例，代码如下：

```
//第 1 章/InputSystemExample.cs

using UnityEngine;
using UnityEngine.InputSystem;

public class InputSystemExample : MonoBehaviour
```

```
{
    private void Update()
    {
        GetKeyInputExample();
    }
    private void GetKeyInputExample()
    {
        var aKey=Keyboard.current.aKey;
        if (aKey.wasPressedThisFrame)
            Debug.Log("键盘 A 键开始被按下");
        if (aKey.isPressed)
            Debug.Log("键盘 A 键持续被按下");
        if (aKey.wasReleasedThisFrame)
            Debug.Log("键盘 A 键被抬起");
    }
}
```

运行程序,按下键盘上的 A 键并抬起,结果如图 1-19 所示。

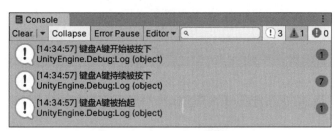

图 1-19 获取键盘 A 键输入

2. 获取鼠标按键输入

Mouse 类是 Pointer 类的派生类,而 Pointer 类与 Keyboard 类一样,是 InputDevice 类的派生类,表示指示器类型输入设备,例如鼠标、触摸屏等。通过 Pointer 类中的 position 属性可以获取光标的位置,通过 delta 属性可以获取光标移动产生的偏移量。

Mouse 类中包含鼠标设备各按键对应的按键控制类成员属性,详情见表 1-7。

表 1-7 鼠标按键对应关系表

属 性	鼠 标 按 键	属 性	鼠 标 按 键
leftButton	鼠标左键	backButton	后退按键
middleButton	滚动按键	forwardButton	前进按键
rightButton	鼠标右键		

这些成员属性同样是 ButtonControl 类型,因此使用方法与键盘按键一样,以获取鼠标

左键输入为例,代码如下:

```
//第 1 章/InputSystemExample.cs

private void Update()
{
    GetMouseInputExample();
}
private void GetMouseInputExample()
{
    var leftButton=Mouse.current.leftButton;
    if (leftButton.wasPressedThisFrame)
        Debug.Log("鼠标左键开始被按下");
    if (leftButton.isPressed)
        Debug.Log("鼠标左键持续被按下");
    if (leftButton.wasReleasedThisFrame)
        Debug.Log("鼠标左键被抬起");
}
```

运行程序,按下鼠标左键并抬起,结果如图1-20所示。

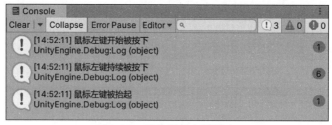

图 1-20 获取鼠标左键输入

3. 获取触摸屏输入

表示触摸屏类型输入设备的是 Touchscreen 类,在旧输入系统中,通过调用 Input.GetTouch(0)获取首要触摸输入,而在新输入系统中,通过 Touchscreen 类中的 primaryTouch 属性即可获取首要触摸输入,表示所有触摸输入的集合通过 touchs 属性获取,示例代码如下:

```
//第 1 章/InputSystemExample.cs

private void Update()
{
    GetTouchScreenInputExample();
}
private void GetTouchScreenInputExample()
```

```
{
    if (Touchscreen.current != null
        && Touchscreen.current.primaryTouch != null)
        Debug.Log(string.Format(
            "首要触摸输入坐标:{0}",
            Touchscreen.current.primaryTouch.position.ReadValue()));
}
```

在新输入系统中包含一个用于输入调试的编辑器窗口工具,通过 Window/Analysis/Input Debugger 菜单打开,使用该工具可以通过鼠标模拟触摸输入,选项在窗口左上角的 Options 按钮打开的下拉列表中,如图 1-21 所示。

选项开启后,运行程序,单击鼠标模拟触摸输入,结果如图 1-22 所示。

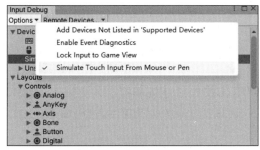

图 1-21　Input Debugger

图 1-22　模拟触摸输入

4. 新旧输入系统的兼容

假设要将某个功能封装为一个 SDK,以便提供给第三方使用,在这个功能中需要获取用户的输入,而第三方既可能使用旧输入系统,也可能使用新输入系统,因此在该功能实现中获取用户输入时,需要兼顾新旧两套输入系统。

实现兼容可以通过宏 ENABLE_INPUT_SYSTEM 判断是否使用新输入系统,如果未使用,则调用旧输入系统的 API,如果使用,则调用新输入系统的 API。

以获取按键输入状态为例,首先定义按键的输入状态类型,PRESSED 表示按键开始被按下,HOLD 表示按键持续被按下,RELEASED 表示按键被抬起,NONE 表示默认状态,代码如下:

```
//第 1 章/KeyState.cs

//<summary>
//按键状态
//</summary>
public enum KeyState
{
    NONE,
```

```
        //<summary>
        //按键开始被按下
        //</summary>
        PRESSED,
        //<summary>
        //按键持续被按下
        //</summary>
        HOLD,
        //<summary>
        //按键被抬起
        //</summary>
        RELEASED,
}
```

在旧输入系统中,键盘按键通过 KeyCode 枚举值表示,在新输入系统中,键盘按键通过 Key 枚举值表示,Keyboard 类内部通过 Key 枚举值获取指定的按键控制。大部分 KeyCode 表示的键盘按键在 Key 中有对应的实现。

在 InputDevice 的基类 InputControl 中,包含用于获取指定按键控制的方法 GetChildControl(),实现兼容时,可以通过该方法获取与 KeyCode 对应的按键控制,代码如下:

```
//第1章/InputUtility.cs

using UnityEngine;

#if ENABLE_INPUT_SYSTEM
using UnityEngine.InputSystem;
using UnityEngine.InputSystem.Controls;
#endif

public class InputUtility
{
    //<summary>
    //获取指定按键的状态
    //</summary>
    //<param name="keyCode">按键</param>
    //<returns>按键状态</returns>
    public static KeyState GetKeyState(KeyCode keyCode)
    {
        if (keyCode ==KeyCode.None) return KeyState.NONE;
//未使用新输入系统
#if !ENABLE_INPUT_SYSTEM
```

```
            return Input.GetKeyDown(keyCode) ? KeyState.PRESSED
                : Input.GetKey(keyCode) ? KeyState.HOLD
                : Input.GetKeyUp(keyCode) ? KeyState.RELEASED
                : KeyState.NONE;
//使用新输入系统
#else
            ButtonControl buttonCtrl;
            int keyCodeV=(int)keyCode;
            //字母 A~Z
            if (keyCodeV>=(int)KeyCode.A
                && keyCodeV <=(int)KeyCode.Z)
            {
                buttonCtrl=Keyboard.current.GetChildControl<KeyControl>(
                    keyCode.ToString());
            }
            //数字 0~9
            else if (keyCodeV>=(int)KeyCode.Alpha0
                && keyCodeV <=(int)KeyCode.Alpha9)
            {
                buttonCtrl=Keyboard.current.GetChildControl<KeyControl>(
                    (keyCodeV-(int)KeyCode.Alpha0).ToString());
            }
            //数字小键盘 0~9
            else if (keyCodeV>=(int)KeyCode.Keypad0
                && keyCodeV <=(int)KeyCode.Keypad9)
            {
                buttonCtrl=Keyboard.current.GetChildControl<KeyControl>(
                    string.Format("numpad{0}",
                        keyCodeV-(int)KeyCode.Keypad0));
            }
            //F1~F15
            else if (keyCodeV>=(int)KeyCode.F1
                && keyCodeV <=(int)KeyCode.F15)
            {
                buttonCtrl=Keyboard.current.GetChildControl<KeyControl>(
                    keyCode.ToString());
            }
            else
            {
                switch (keyCode)
                {
```

```
        case KeyCode.Backspace:
            buttonCtrl=Keyboard.current.backspaceKey; break;
        case KeyCode.Delete:
            buttonCtrl=Keyboard.current.deleteKey; break;
        case KeyCode.Tab:
            buttonCtrl=Keyboard.current.tabKey; break;
        case KeyCode.KeypadPeriod:
            buttonCtrl=Keyboard.current.numpadPeriodKey; break;
        case KeyCode.KeypadDivide:
            buttonCtrl=Keyboard.current.numpadDivideKey; break;
        case KeyCode.KeypadMultiply:
            buttonCtrl=Keyboard.current.numpadMultiplyKey; break;
        case KeyCode.KeypadMinus:
            buttonCtrl=Keyboard.current.numpadMinusKey; break;
        case KeyCode.KeypadPlus:
            buttonCtrl=Keyboard.current.numpadPlusKey; break;
        case KeyCode.KeypadEnter:
            buttonCtrl=Keyboard.current.numpadEnterKey; break;
        case KeyCode.KeypadEquals:
            buttonCtrl=Keyboard.current.numpadEqualsKey; break;
        case KeyCode.UpArrow:
            buttonCtrl=Keyboard.current.upArrowKey; break;
        case KeyCode.DownArrow:
            buttonCtrl=Keyboard.current.downArrowKey; break;
        case KeyCode.RightArrow:
            buttonCtrl=Keyboard.current.rightArrowKey; break;
        case KeyCode.LeftArrow:
            buttonCtrl=Keyboard.current.leftArrowKey; break;
        case KeyCode.Insert:
            buttonCtrl=Keyboard.current.insertKey; break;
        case KeyCode.Home:
            buttonCtrl=Keyboard.current.homeKey; break;
        case KeyCode.End:
            buttonCtrl=Keyboard.current.endKey; break;
        case KeyCode.PageUp:
            buttonCtrl=Keyboard.current.pageUpKey; break;
        case KeyCode.PageDown:
            buttonCtrl=Keyboard.current.pageDownKey; break;
        case KeyCode.Quote:
            buttonCtrl=Keyboard.current.quoteKey; break;
        case KeyCode.Comma:
```

```
                buttonCtrl=Keyboard.current.commaKey; break;
            case KeyCode.Minus:
                buttonCtrl=Keyboard.current.minusKey; break;
            case KeyCode.Period:
                buttonCtrl=Keyboard.current.periodKey; break;
            case KeyCode.Slash:
                buttonCtrl=Keyboard.current.slashKey; break;
            case KeyCode.Semicolon:
                buttonCtrl=Keyboard.current.semicolonKey; break;
            case KeyCode.BackQuote:
                buttonCtrl=Keyboard.current.backquoteKey; break;
            case KeyCode.Numlock:
                buttonCtrl=Keyboard.current.numLockKey; break;
            case KeyCode.CapsLock:
                buttonCtrl=Keyboard.current.capsLockKey; break;
            case KeyCode.ScrollLock:
                buttonCtrl=Keyboard.current.scrollLockKey; break;
            case KeyCode.LeftShift:
                buttonCtrl=Keyboard.current.leftShiftKey; break;
            case KeyCode.RightShift:
                buttonCtrl=Keyboard.current.rightShiftKey; break;
            case KeyCode.LeftControl:
                buttonCtrl=Keyboard.current.leftCtrlKey; break;
            case KeyCode.RightControl:
                buttonCtrl=Keyboard.current.rightCtrlKey; break;
            case KeyCode.LeftAlt:
                buttonCtrl=Keyboard.current.leftAltKey; break;
            case KeyCode.RightAlt:
                buttonCtrl=Keyboard.current.rightAltKey; break;
            case KeyCode.LeftCommand:
                buttonCtrl=Keyboard.current.leftCommandKey; break;
            case KeyCode.RightCommand:
                buttonCtrl=Keyboard.current.rightCommandKey; break;
            case KeyCode.LeftWindows:
                buttonCtrl=Keyboard.current.leftWindowsKey; break;
            case KeyCode.RightWindows:
                buttonCtrl=Keyboard.current.rightWindowsKey; break;
            case KeyCode.Print:
                buttonCtrl=Keyboard.current.printScreenKey; break;
            default: buttonCtrl=null; break;
    }
```

```
            }
            return buttonCtrl !=null
                ? (buttonCtrl.wasPressedThisFrame ? KeyState.PRESSED
                    : buttonCtrl.isPressed ? KeyState.HOLD
                    : buttonCtrl.wasReleasedThisFrame ? KeyState.RELEASED
                    : KeyState.NONE)
                : KeyState.NONE;
#endif
    }
}
```

有了上述方法,在获取键盘按键输入时无须再关注使用的是旧输入系统还是新输入系统,通过获取按键状态便可以得知按键是否被按下或抬起,示例代码如下:

```
//第1章/InputSystemExample.cs

private void Update()
{
    InputUtilityExample();
}
private void InputUtilityExample()
{
    KeyState keyState=InputUtility.GetKeyState(KeyCode.A);
    if (keyState ==KeyState.PRESSED)
        Debug.Log("键盘A键开始被按下");
    else if (keyState ==KeyState.HOLD)
        Debug.Log("键盘A键持续被按下");
    else if (keyState ==KeyState.RELEASED)
        Debug.Log("键盘A键被抬起");
}
```

以上是获取键盘按键输入的示例,在获取鼠标设备的输入时,例如获取左键的单击位置,在旧输入系统中需要调用 Input 类中的方法判断鼠标左键是否被单击并通过 mousePosition 属性获取鼠标位置,在新输入系统中,则需要通过 Mouse 类中的属性获取鼠标左键是否被按下及鼠标位置信息,兼容方法的代码如下:

```
//第1章/InputUtility.cs

//<summary>
//鼠标左键是否被单击
//</summary>
//<param name="clickPosition">单击的位置</param>
```

```
//<returns>是否被单击</returns>
public static bool IsLeftMouseButtonClick(out Vector2 clickPosition)
{
    bool isClick=false;
    clickPosition=Vector2.zero;
#if !ENABLE_INPUT_SYSTEM
    isClick=Input.GetMouseButtonDown(0);
#else
    isClick=Mouse.current !=null
        && Mouse.current.leftButton.wasPressedThisFrame;
#endif
    if (isClick)
    {
#if !ENABLE_INPUT_SYSTEM
        clickPosition=Input.mousePosition;
#else
        clickPosition=Mouse.current.position.ReadValue();
#endif
    }
    return isClick;
}
```

1.2.2 Input Action Asset 配置文件

1.2.1 节介绍了新输入系统的简单使用方式，它可以像旧输入系统一样使用，但这种使用方式显然体现不出新输入系统的任何优势。

如果目标平台并不是单一的，则在获取输入时就要兼顾各平台。例如，在获取屏幕单击位置时，在 PC 端需要通过 Mouse 类的相关属性获取，在移动端需要通过 Touchscreen 类的相关属性获取，示例代码如下：

```
//第 1 章/InputSystemExample.cs

private bool IsClickExample(out Vector3 clickPosition)
{
    bool isClick=false;
    clickPosition=Vector3.zero;
#if UNITY_STANDALONE
    isClick=Mouse.current !=null
        && Mouse.current.leftButton.wasPressedThisFrame;
#elif UNITY_ANDROID
    isClick=Touchscreen.current !=null
```

```
            && Touchscreen.current.primaryTouch !=null;
#endif
    if (isClick)
    {
#if UNITY_STANDALONE
        clickPosition=Mouse.current.position.ReadValue();
#elif UNITY_ANDROID
        clickPosition=Touchscreen.current.primaryTouch
            .position.ReadValue();
#endif
    }
    return isClick;
}
```

可以看到，为了兼顾不同平台的输入，代码书写非常烦琐，而新输入系统中的 Input Action Asset 可以很好地解决该问题，它作为一种配置文件，存储于项目工程中，通过菜单 Assets/Create/ Input Actions 可以创建该类型的资产。

双击打开配置文件，当未进行任何配置时，默认如图 1-23 所示。左侧的 Action Maps 表示该配置文件的分组，每个 Input Action Map 中包含一个 Input Action 的数组。

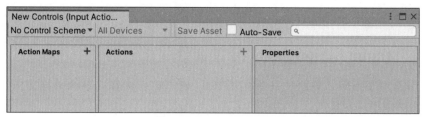

图 1-23　Input Actions

以获取屏幕单击位置为例，创建一个新的 Action Map，并在该 Action Map 中创建一个新的 Input Action，命名为 IsClick，如图 1-24 所示。

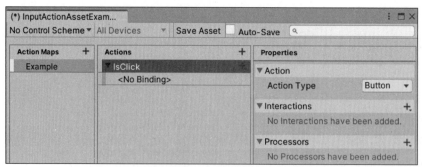

图 1-24　创建 IsClick

在 Input Action 展开的折叠栏中列举这个 Input Action 中包含的 Input Control 列表，默认显示 No Binding，表示没有绑定任何设备的输入，选中并在右侧 Properties 窗口中进行绑定，如图 1-25 所示，可以绑定多种设备的输入。

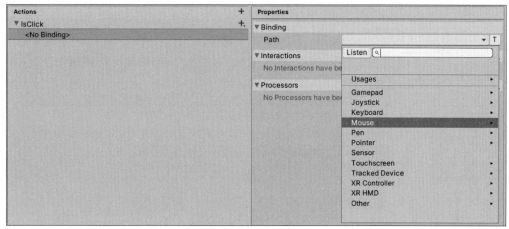

图 1-25　Input Control

假设单击事件既可以通过鼠标单击触发，也可以通过触摸屏单点触摸触发，那么需要为 IsClick 创建两个 Input Control，分别绑定 Mouse 和 Touchscreen 中的 Press 选项，如图 1-26 所示。

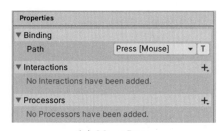 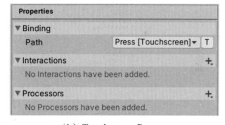

（a）Mouse Press　　　　　　　（b）Touchscreen Press

图 1-26　Input Control Binding

再创建一个 Input Action，命名为 MousePosition，用于获取单击位置，Action Type 为 Value 类型，Control Type 为 Vector 2 类型。在这个 Input Action 中创建两个 Input Control，分别绑定 Mouse 中的 Position 选项和 Touchscreen 中 Primary Touch 的 Position 选项，如图 1-27 所示。

编辑好配置文件后，选中该配置文件，在其检视面板中勾选 Generate C♯ Class 并应用便可以生成对应的 C♯ 类脚本文件，如图 1-28 所示。

有了该类后，获取屏幕单击位置的代码便可以以简便的方式书写，可以看到，不再需要书写判断运行平台的烦琐代码，代码如下：

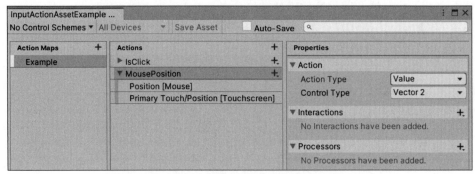

图 1-27　创建 MousePosition

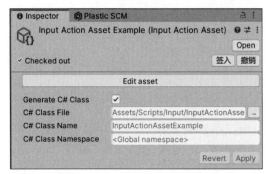

图 1-28　Generate C#Class

```
//第 1 章/InputSystemExample.cs

private InputActionAssetExample actionAsset;

private void Start()
{
    actionAsset=new InputActionAssetExample();
    actionAsset.Enable();
}
private void Update()
{
    if (actionAsset.Example.IsClick.phase ==InputActionPhase.Started)
        Debug.Log(actionAsset.Example.MousePosition.ReadValue<Vector2>());
}
```

1.2.3　Player Input 组件

在 1.2.2 节中介绍了配置 Input Action Asset 文件并通过其生成的脚本类获取输入的方法，该脚本类的生成并非是必需的，也可以通过 Player Input 组件添加事件监听的方式获取输入，该组件的检视面板如图 1-29 所示。

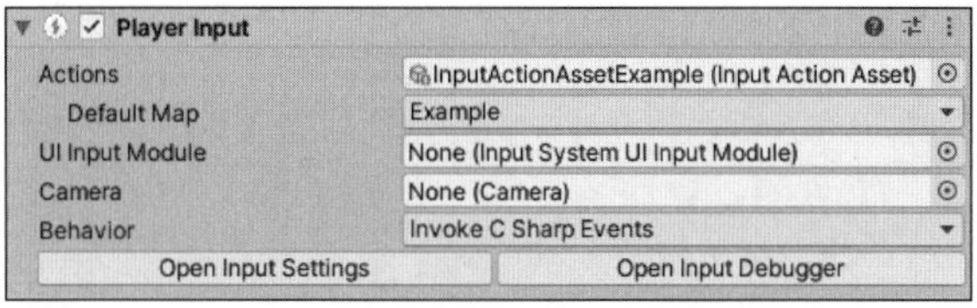

图 1-29　Player Input 组件

事件监听的方式通过 Behavior 进行设置，有 4 种类型，详解见表 1-8。

表 1-8　Player Input Behavior

类　　型	详　　解
Send Messages	事件逻辑组件挂载于 Player Input 组件所在的物体，通过发送消息的方式通知执行事件
Broadcast Messages	事件逻辑组件挂载于 Player Input 组件所在的物体或其子物体，通过广播消息的方式通知执行事件
Invoke Unity Events	通过在检视面板拖曳关联 Unity Event 指定要执行的回调事件
Invoke C Sharp Events	通过 C Sharp 事件委托添加回调事件

Send Messages 和 Broad Messages 这两种类型是通过反射的方式调用事件的，事件在脚本中以 On 拼接 Input Action 的名称命名。由于使用反射的效率低下，因此通常推荐使用 Invoke Unity Events 或 Invoke C Sharp Events 类型。

以 Invoke C Sharp Events 类型为例，在脚本中获取 Player Input 组件对象，并为其添加回调事件，代码如下：

```
//第 1 章/InputSystemExample.cs

private PlayerInput playerInput;

private void OnEnable()
{
    if (playerInput ==null)
    {
        playerInput=GetComponentInChildren<PlayerInput>();
        playerInput.currentActionMap.Enable();
    }
    playerInput.onActionTriggered +=OnActionTriggered;
}
private void OnDisable()
{
```

```
        if (playerInput !=null)
            playerInput.onActionTriggered-=OnActionTriggered;
}
private void OnActionTriggered(InputAction.CallbackContext context)
{
    switch (context.action.name)
    {
        case "IsClick":
            Debug.Log("单击");
            break;
        case "MousePosition":
            Debug.Log(string.Format("鼠标位置:{0}",
                context.ReadValue<Vector2>()));
            break;
        default:
            break;
    }
}
```

第 2 章 数 学 基 础

在游戏开发中,数学基础是不可或缺的一部分,无论是构建复杂的物理系统,还是实现逼真的动画效果都需要对基本的数学概念和运算有深入的理解。本章将重点介绍 Unity 开发中常用的数学基础知识,包括 Mathf 数学运算工具类、向量的运算及矩阵的运算,帮助读者建立起坚实的数学基础,为后续的游戏开发打下坚实的基础。

2.1 Mathf

Mathf 类是 Unity 提供的一个用于数学运算的工具类,它包含一系列常用的数学函数和常量,可以大大地简化开发过程中的数学计算。本节将详细介绍 Mathf 类的基本用法,包括常见的数学函数,如三角函数、插值函数等,以及常用的数学常量。通过学习和掌握 Mathf 类的使用,读者可以更加高效地进行数学计算,提高开发效率。

2.1.1 常量

在学习 Mathf 类中的数学运算函数之前,先来了解 Mathf 类中包含的重要常量,详解见表 2-1。

表 2-1 Mathf 类中的常量

常　　量	详　　解
PI	圆周率,即圆的周长与直径的比值,约等于 3.141592654
Infinity	正无穷大的表示形式
NegativeInfinity	负无穷大的表示形式
Deg2Rad	度转弧度的换算常量,值等于 PI/180
Rad2Deg	弧度转度的换算常量,值等于 180/PI

一些常量经常会参与到数学运算中,例如,在求 60°的余弦值时,需要使用 Cos()函数,该函数的参数以弧度为单位,因此需要通过 Deg2Rad 常量将度换算为弧度再将结果作为参

数传入，示例代码如下：

```
Debug.Log(Mathf.Cos(60f * Mathf.Deg2Rad)); //0.5
```

除了以上常量外，Mathf 类中还包含一个静态只读的 float 类型字段 Epsilon，它表示微小浮点值，也就是大于 0 的最小浮点值。

2.1.2 三角函数

Mathf 类中提供了正弦、余弦、正切三角函数，以及反正弦、反余弦和反正切反三角函数，详解见表 2-2。

表 2-2　Mathf 类中的三角函数

三角函数	详解
Sin(float f)	返回角度 f 的正弦值，f 以弧度为单位
Cos(float f)	返回角度 f 的余弦值，f 以弧度为单位
Tan(float f)	返回角度 f 的正切值，f 以弧度为单位
Asin(float f)	返回 f 的反正弦值
Acos(float f)	返回 f 的反余弦值
Atan(float f)	返回 f 的反正切值
Atan2(float y, float x)	返回正切为 y/x 的角度

需要注意，以上反三角函数返回的角度值以弧度为单位，如果要换算为度，则可以使用常量 Rad2Deg 进行换算，示例代码如下：

```
Debug.Log(Mathf.Acos(0.5f) * Mathf.Rad2Deg); //60
```

2.1.3 插值函数

插值函数可以用于计算两个给定点之间某处的值，Mathf 类中提供了一系列插值函数，它们具有不同的行为，适用于不同的场景，详解见表 2-3。

表 2-3　Mathf 类中的插值函数

插值函数	详解
Lerp(float a, float b, float t)	在 a 与 b 之间根据 t 进行线性插值，t 的取值范围为[0, 1]
LerpAngle(float a, float b, float t)	与 Lerp() 相同，但是用于处理角度
LerpUnclamped(float a, float b, float t)	在 a 与 b 之间根据 t 进行线性插值，t 没有取值范围限制
MoveTowards(float current, float target, float maxDelta)	根据 maxDelta 从当前值向目标值移动

续表

插值函数	详解
MoveTowardsAngle(float current, float target, float maxDelta)	与 MoveTowards() 相同，但是用于处理角度
SmoothDamp(float current, float target, ref float currentVelocity, float smoothTime)	根据 currentVelocity 和 smoothTime 从当前值向目标值进行平滑插值，currentVelocity 表示当前速度，smoothTime 表示达到目标值所需的近似时间
SmoothDampAngle(float current, float target, ref float currentVelocity, float smoothTime)	与 SmoothDamp() 相同，但是用于处理角度
SmoothStep(float from, float to, float t)	在 from 与 to 之间根据 t 进行插值，但是插值会从起点逐渐加速，然后朝着终点减慢

Lerp() 是最基本、最常用的插值函数，例如用于人物角色移动速度的计算，假设角色行走时的速度为 1，奔跑时的速度为 3，那么从走到跑或者从跑到走的过渡期间，便可以使用该函数计算用于过渡期间的移动速度。

与 Lerp() 函数对应的是 InverseLerp() 函数，它用于获取在范围 $[a, b]$ 内生成插值 value 的线性参数值，代码如下：

```
public static float InverseLerp(float a, float b, float value);
```

2.1.4 最值与限制函数

计算两个或更多值中的最大值、最小值是游戏开发中的常见需求，Mathf 类中提供了这些运算函数，代码如下：

```
public static float Max(float a, float b);
public static float Max(params float[] values);
public static int Max(int a, int b);
public static int Max(params int[] values);
public static float Min(float a, float b);
public static float Min(params float[] values);
public static int Min(int a, int b);
public static int Min(params int[] values);
```

当需要将值限制在某个取值范围时，可以使用限制函数 Clamp()，参数 value 表示要限制在最小值和最大值定义的取值范围内的数值，min 和 max 则分别表示最小值和最大值。Clamp01() 函数不需要指定最小值、最大值，表示将 value 限制在 $[0, 1]$ 取值范围内，代码如下：

```
public static float Clamp(float value, float min, float max);
```

```
public static int Clamp(int value, int min, int max);
public static float Clamp01(float value);
```

除此之外,开发工作中还经常会遇到小数取整的需求,Mathf 类中包含向上取整、向下取整和舍入取整等函数,详解见表 2-4。

表 2-4　Mathf 类中的取整函数

取整函数	详解
Ceil(float f)	返回大于或等于 f 的最小整数,返回值是 float 类型
CeilToInt(float f)	返回大于或等于 f 的最小整数,返回值是 int 类型
Floor(float f)	返回小于或等于 f 的最大整数,返回值是 float 类型
FloorToInt(float f)	返回小于或等于 f 的最大整数,返回值是 int 类型
Round (float f)	返回最接近 f 的整数,返回值是 float 类型
RoundToInt(float f)	返回最接近 f 的整数,返回值是 int 类型

值得注意的是,Round()和 RoundToInt()这两个舍入函数和数学运算中的四舍五入法略有差别。在四舍五入法中,如果尾数的最高位数字是 4 或比 4 小,就把尾数舍去,如果尾数的最高位数字是 5 或比 5 大,就把尾数舍去并且在它的前一位进一,而在 Round()或者 RoundToInt()函数中,如果尾数是 0.5,则函数将返回邻近 f 的两个整数中的偶数。例如,当 f 为 2.5 时,返回值为 2,当 f 为 3.5 时,返回值为 4,示例代码如下:

```
Debug.Log(Mathf.Round(2.5f)); //2
Debug.Log(Mathf.Round(3.5f)); //4
```

2.1.5　幂、平方根、对数函数

Mathf 类中提供了幂运算、开平方运算和对数运算的函数,详解见表 2-5。

表 2-5　Mathf 类中的幂、平方根、对数函数

函　数	详　解
Pow(float f, float p)	返回 f 的 p 次幂
Sqrt(float f)	返回 f 的平方根
Exp(float power)	返回 e 的指定幂
ClosestPowerOfTwo(int value)	返回最接近的 2 的幂值
NextPowerOfTwo(int value)	返回大于或等于 value 的下一个 2 的幂
IsPowerOfTwo(int value)	如果 value 是 2 的幂,则返回值为 true,否则返回值为 false

续表

函 数	详 解
Log(float f,float p)	返回以 p 为底 f 的对数
Log(float f)	返回以无理数 e 为底 f 的对数
Log10(float f)	返回以 10 为底 f 的对数

2.2 向量

向量是 Unity 开发中非常重要的数学概念之一,它是一种同时具有大小和方向的量,又称矢量。与向量对应的量叫作标量(或数量),标量没有方向,只有大小。

向量可以用于表示空间中的方向和大小,是进行游戏物体位置、速度和方向等属性计算的基础。本节将详细讲解向量的基本运算规则,包括向量的加法、减法、数乘、点乘和叉乘等。通过学习和掌握向量的运算规则,读者可以更加灵活地处理游戏中的各种空间关系,实现更加逼真的游戏效果。

向量可以在多个维度上表示,Unity 提供了 Vector2、Vector3 和 Vector4 类,分别用于处理二维、三维和四维向量。

2.2.1 向量加减

当两个向量的维度相同时,两个向量相加等于两个向量对应的分量相加,公式如下:

$$(x_1,y_1)+(x_2,y_2)=(x_1+x_2,y_1+y_2) \qquad (2-1)$$

向量的加法满足平行四边形法则,即两个向量合成时,以表示这两个向量的线段为邻边作平行四边形,该平行四边形的对角线就是合向量的大小和方向,如图 2-1(a)所示。

向量的加法同时满足三角形法则,即两个向量合成时,其合向量应当为将一个向量的起点移动到另一个向量的终点,合向量就是第 1 个向量的起点到第 2 个向量的终点,如图 2-1(b)所示。

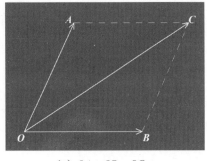

(a) **OA** + **OB** = **OC**

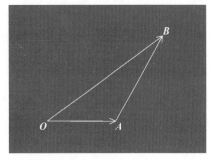

(b) **OA** + **AB** = **OB**

图 2-1 向量加法

两个维度相等的向量相减,等于两个向量对应的分量相减,公式如下:
$$(x_1, y_1) - (x_2, y_2) = (x_1 - x_2, y_1 - y_2) \quad (2-2)$$

向量的减法可以理解为加上一个向量的相反向量,也就是说,向量减法是向量加法的逆运算。把两个向量的起点放到一个共同起点,由一个向量终点引向另一个向量终点的向量就是两者之差向量,箭头指向谁,谁就是被减数向量。

如图 2-2 所示,向量 **OB** 减向量 **OA** 等于从向量 **OA** 的终点指向向量 **OB** 的终点的向量,这是向量减法的几何意义。

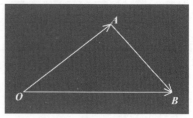

图 2-2 **OB** − **OA** = **AB**

2.2.2 向量数乘

向量具有大小和方向,而标量只有大小,将向量乘以标量时,如果标量值大于 0,则产生的向量与原始向量具有相同的方向,但是大小等于原始向量的大小乘以标量值,如果标量值小于 0,则产生的向量与原始向量方向相反,公式如下:
$$k(x, y) = (kx, ky) \quad (2-3)$$

向量数乘的几何意义就是通过改变向量的长度和方向实现向量的缩放和旋转。当任何向量除以其自身的大小时,得到的结果是大小为 1 的向量,即所谓的归一化向量。如果归一化向量乘以标量,则结果的大小等于该标量值。

2.2.3 向量插值

Unity 提供的向量类中提供了向量插值运算函数 Lerp(),以 Vector3 类为例,插值函数的代码如下,参数 a 表示起始值,b 表示结束值,函数会根据参数 t 在起始值和结束值之间进行线性插值,参数 t 的取值范围为[0,1],当 t 为 0 时,函数返回 a,当 t 为 1 时,函数返回 b,当 t 为 0.5 时,函数返回 a 和 b 的中点。

```
public static Vector3 Lerp (Vector3 a, Vector3 b, float t);
```

Vector3 类中除了 Lerp()函数外,还包含另一个插值函数 Slerp(),两者的区别在于,前者用于进行线性插值,而后者用于进行球形插值,如图 2-3 所示。

2.2.4 向量点乘与叉乘

向量点乘又叫向量的点积、内积,以三维向量为例,两个三维向量的点乘就是将两个向量的对应分量相乘后再求和,结果是一个标量,公式如下:
$$(x_1, y_1, z_1) \cdot (x_2, y_2, z_2) = x_1 x_2 + y_1 y_2 + z_1 z_2 \quad (2-4)$$

向量的点乘满足交换律,即 $a \cdot b = b \cdot a$。

从几何的角度看,两个向量点乘等于两个向量大小相乘并乘以两个向量之间夹角余弦

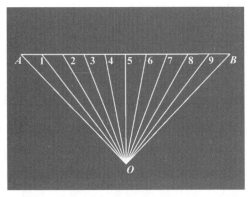 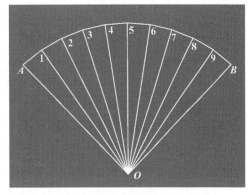

（a）线性插值　　　　　　　　　　　　（b）球形插值

图 2-3　线性插值与球形插值

值的结果，公式如下，因此，向量点乘可以看作一个向量和它在另一个向量上的投影的长度的乘积。

$$\boldsymbol{a} \cdot \boldsymbol{b} = |\boldsymbol{a}||\boldsymbol{b}|\cos\theta \tag{2-5}$$

如果两个向量均是归一化向量，则它们的模长均为1，此时两个向量的点乘结果就是两个向量之间夹角的余弦值，因此通过向量点乘可以进而求得两个向量之间的夹角。

这十分有用，例如，如果要判断物体 B 在物体 A 的前方还是后方，则可以通过计算物体 A 的前方向量和物体 A 指向物体 B 的方向的归一化向量的点乘结果，如果点乘结果大于 0，则说明两个向量之间的夹角小于 90°，那么物体 B 在物体 A 的前方，如图 2-4(a)所示，如果点乘结果小于 0，则说明两个向量之间的夹角大于 90°，那么物体 B 在物体 A 的后方，如图 2-4(b)所示。

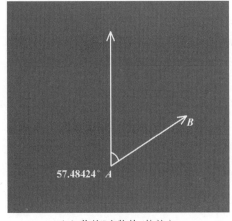 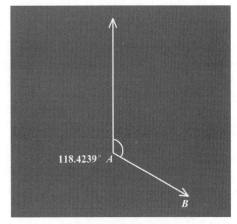

（a）物体B在物体A的前方　　　　　　　（b）物体B在物体A的后方

图 2-4　物体 A 与物体 B 的前后关系

向量叉乘又叫向量的叉积、外积，与点乘不同的是，叉乘的结果不是标量，而是一个新的向量,这个新的向量垂直于两个原始向量所在的平面,计算公式如下。

$$(x_1, y_1, z_1) \times (x_2, y_2, z_2) = (y_1 z_2 - z_1 y_2, z_1 x_2 - x_1 z_2, x_1 y_2 - y_1 x_2) \quad (2\text{-}6)$$

需要注意的是,叉乘不满足交换律,即 $a \times b \neq b \times a$。由于 Unity 中使用的是左手坐标系,因此两个向量传递到叉乘函数的顺序可以通过左手法则确定。

从几何的角度看,向量叉乘得到的向量等于两个原始向量模长的乘积并乘以它们之间夹角的正弦值,公式如下。

$$a \times b = |a||b|\sin\theta \quad (2\text{-}7)$$

通过点乘可以判断物体 B 在物体 A 的前方还是后方,而叉乘可以判断物体 B 在物体 A 的左方还是右方,计算物体 A 的前方向量和物体 A 指向物体 B 的方向的归一化向量的叉乘结果,如果结果的 y 值大于 0,则物体 B 在物体 A 的右侧,如图 2-5(a)所示,如果 y 值小于 0,则物体 B 在物体 A 的左侧,如图 2-5(b)所示。

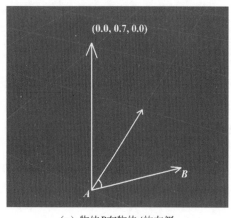
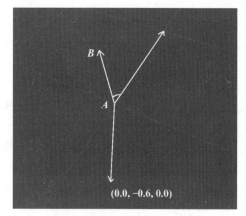

(a) 物体 B 在物体 A 的右侧　　　　　　(b) 物体 B 在物体 A 的左侧

图 2-5　物体 A 与物体 B 的左右关系

通过向量点乘和叉乘判断方向关系的方法被广泛地应用于游戏开发中,例如,游戏单位 A 在其正前方施放一个攻击范围为扇形区域的技能,如何判断游戏单位 B 是否被该技能命中? 需要满足以下两个条件:

(1) 游戏单位 A 的正前方与游戏单位 A 指向游戏单位 B 的方向之间的夹角小于或等于扇形角度的二分之一。

(2) 游戏单位 A 指向游戏单位 B 的方向向量的模长小于或等于扇形的半径。

Vector3 类中用于点乘和叉乘运算的函数分别为 Dot()和 Cross(),是否满足条件(1)便需要用到 Dot()函数,代码如下:

```
//第 2 章/SectorAttack.cs

using UnityEngine;
```

```csharp
public class SectorAttack
{
    /// <summary>
    ///检测目标是否被扇形攻击命中
    /// </summary>
    /// <param name="sectorAngle">扇形的角度</param>
    /// <param name="sectorRadius">扇形的半径</param>
    /// <param name="attacker">攻击者</param>
    /// <param name="checkTarget">检测目标</param>
    /// <returns></returns>
    public static bool IsInRange(float sectorAngle, float sectorRadius,
        Transform attacker, Transform checkTarget)
    {
        //攻击者指向检测目标的方向
        Vector3 direction =checkTarget.position
            -attacker.position;
        //点乘得到两个归一化向量之间的余弦值
        float dot =Vector3.Dot(attacker.forward, direction.normalized);
        //通过反余弦函数计算两个向量的夹角
        float angle =Mathf.Acos(dot) * Mathf.Rad2Deg;
        return angle <=sectorAngle * .5f
&& direction.magnitude <=sectorRadius;
    }
}
```

2.3 矩阵

矩阵是另一个在游戏开发中非常重要的数学概念。它可以用来表示物体的变换,如平移、旋转和缩放,是进行复杂物理计算和图形渲染的基础。本节将介绍矩阵的基本概念和运算规则,包括矩阵的表示方法、乘法、转置和逆等。同时,本节还将介绍 Unity 中常用的变换矩阵,即平移矩阵、缩放矩阵和旋转矩阵,利用这些矩阵可以进行物体的变换操作。通过学习和掌握矩阵的运算规则,读者可以更加精确地控制游戏物体的位置和姿态,实现更加丰富的游戏效果。

矩阵是描述线性变换的一种数学工具,由 $m \times n$ 个数排列成的 m 行 n 列的数表称为 m 行 n 列的矩阵,简称 $m \times n$ 矩阵。记作:

$$\boldsymbol{A} = \begin{bmatrix} a_{11} & a_{12} & \cdots & a_{1n} \\ a_{21} & a_{22} & \cdots & a_{2n} \\ \cdots & \cdots & \ddots & \cdots \\ a_{m1} & a_{m2} & \cdots & a_{mn} \end{bmatrix}$$

矩阵中的数称为元素，矩阵元素通过行和列的双下标进行标识，A_{ij} 表示位于矩阵 A 的第 i 行第 j 列的元素。如果矩阵的行数和列数相等，都为 n，则该矩阵称为 n 阶矩阵或 n 阶方阵。Unity 中提供了用于处理矩阵的类 Matrix4x4，它代表的便是 4 阶方阵。

将矩阵 A 的行和列互相交换所产生的矩阵称为 A 的转置矩阵 A^T，这一过程称为矩阵的转置。由转置矩阵的定义可知，一个 $m \times n$ 的矩阵的转置矩阵是一个 $n \times m$ 的矩阵，例如

$$\begin{bmatrix} 1 & 5 & 2 \\ 0 & -3 & 4 \end{bmatrix}^T = \begin{bmatrix} 1 & 0 \\ 5 & -3 \\ 2 & 4 \end{bmatrix}$$

如果 n 阶方阵和它的转置矩阵相等，即 $A^T = A$，则称 A 矩阵为对称矩阵。

如下所示，如果 n 阶方阵从左上角到右下角的对角线上的元素均为 1，其他元素均为 0，则称该矩阵为单位矩阵，通常用 I 或 E 来表示，记为 I_n 或 E_n。单位矩阵在矩阵的乘法中起着特殊的作用，任何矩阵和单位矩阵相乘都等于它本身。

$$E_2 = \begin{vmatrix} 1 & 0 \\ 0 & 1 \end{vmatrix} \quad E_3 = \begin{vmatrix} 1 & 0 & 0 \\ 0 & 1 & 0 \\ 0 & 0 & 1 \end{vmatrix} \quad E_n = \begin{vmatrix} 1 & 0 & \cdots & 0 \\ 0 & 1 & \cdots & 0 \\ \cdots & \cdots & \ddots & \cdots \\ 0 & 0 & \cdots & 1 \end{vmatrix}$$

假设矩阵 A 是一个 n 阶方阵，如果存在另一个 n 阶方阵 B，使 $AB = BA = E$，则称矩阵 A 可逆，并称矩阵 B 是矩阵 A 的逆矩阵。单位矩阵的逆矩阵是它本身。

2.3.1 矩阵的基本运算

矩阵的加法或减法就是两个矩阵的相同位置上的元素相加或相减，只有当两个矩阵的行数和列数都相同时，才能进行加减法运算，以 3×3 矩阵为例，公式如下：

$$\begin{vmatrix} a_{11} & a_{12} & a_{13} \\ a_{21} & a_{22} & a_{23} \\ a_{31} & a_{32} & a_{33} \end{vmatrix} + \begin{vmatrix} b_{11} & b_{12} & b_{13} \\ b_{21} & b_{22} & b_{23} \\ b_{31} & b_{32} & b_{33} \end{vmatrix} = \begin{vmatrix} a_{11}+b_{11} & a_{12}+b_{12} & a_{13}+b_{13} \\ a_{21}+b_{21} & a_{22}+b_{22} & a_{23}+b_{23} \\ a_{31}+b_{31} & a_{32}+b_{32} & a_{33}+b_{33} \end{vmatrix} \quad (2\text{-}8)$$

$$\begin{vmatrix} a_{11} & a_{12} & a_{13} \\ a_{21} & a_{22} & a_{23} \\ a_{31} & a_{32} & a_{33} \end{vmatrix} - \begin{vmatrix} b_{11} & b_{12} & b_{13} \\ b_{21} & b_{22} & b_{23} \\ b_{31} & b_{32} & b_{33} \end{vmatrix} = \begin{vmatrix} a_{11}-b_{11} & a_{12}-b_{12} & a_{13}-b_{13} \\ a_{21}-b_{21} & a_{22}-b_{22} & a_{23}-b_{23} \\ a_{31}-b_{31} & a_{32}-b_{32} & a_{33}-b_{33} \end{vmatrix} \quad (2\text{-}9)$$

例如，有 A、B 两个矩阵，如下所示。

$$A = \begin{bmatrix} 12 & 3 & -5 \\ 1 & -9 & 0 \\ 3 & 6 & 8 \end{bmatrix} \quad B = \begin{bmatrix} 1 & 8 & 9 \\ 6 & 5 & 4 \\ 3 & 2 & 1 \end{bmatrix}$$

那么 $A + B$、$A - B$ 分别等于：

$$A + B = \begin{bmatrix} 12+1 & 3+8 & -5+9 \\ 1+6 & -9+5 & 0+4 \\ 3+3 & 6+2 & 8+1 \end{bmatrix} = \begin{bmatrix} 13 & 11 & 4 \\ 7 & -4 & 4 \\ 6 & 8 & 9 \end{bmatrix}$$

$$A - B = \begin{bmatrix} 12-1 & 3-8 & -5-9 \\ 1-6 & -9-5 & 0-4 \\ 3-3 & 6-2 & 8-1 \end{bmatrix} = \begin{bmatrix} 11 & -5 & -14 \\ -5 & -14 & -4 \\ 0 & 4 & 7 \end{bmatrix}$$

矩阵的加减法运算满足交换律、结合律和分配律。

矩阵的加减法和矩阵的数乘合称矩阵的线性运算，矩阵的数乘就是将一个标量和矩阵中的每个元素相乘，以 3×3 矩阵为例，公式如下：

$$k \begin{vmatrix} a_{11} & a_{12} & a_{13} \\ a_{21} & a_{22} & a_{23} \\ a_{31} & a_{32} & a_{33} \end{vmatrix} = \begin{vmatrix} a_{11} & a_{12} & a_{13} \\ a_{21} & a_{22} & a_{23} \\ a_{31} & a_{32} & a_{33} \end{vmatrix} k = \begin{vmatrix} ka_{11} & ka_{12} & ka_{13} \\ ka_{21} & ka_{22} & ka_{23} \\ ka_{31} & ka_{32} & ka_{33} \end{vmatrix} \quad (2\text{-}10)$$

例如将矩阵 A 乘以 2，结果如下：

$$2A = 2\begin{bmatrix} 12 & 3 & -5 \\ 1 & -9 & 0 \\ 3 & 6 & 8 \end{bmatrix} = \begin{bmatrix} 12*2 & 3*2 & -5*2 \\ 1*2 & -9*2 & 0*2 \\ 3*2 & 6*2 & 8*2 \end{bmatrix} = \begin{bmatrix} 24 & 6 & -10 \\ 2 & -18 & 0 \\ 6 & 12 & 16 \end{bmatrix}$$

两个矩阵的乘法仅当第 1 个矩阵 A 的列数和第 2 个矩阵 B 的行数相等时才能定义。如 A 是 $m \times n$ 矩阵，B 是 $n \times p$ 矩阵，它们的乘积 C 是一个 $m \times p$ 矩阵，C_{ij} 的值等于矩阵 A 的第 i 行所对应的行向量与矩阵 B 的第 j 列所对应的列向量的点乘结果，公式如下：

$$C_{i,j} = A_{i,1} B_{1,j} + A_{i,2} B_{2,j} + \cdots + A_{i,n} B_{n,j} = \sum_{r=1}^{n} A_{i,r} B_{r,j} \quad (2\text{-}11)$$

例如

$$\begin{bmatrix} 2 & 0 & 4 \\ -2 & 6 & 2 \end{bmatrix} * \begin{bmatrix} 6 & 2 \\ 4 & 2 \\ 2 & 0 \end{bmatrix} = \begin{bmatrix} 2*6+0*4+4*2 & 2*2+0*2+4*0 \\ -2*6+6*4+2*2 & -2*2+6*2+2*0 \end{bmatrix} = \begin{bmatrix} 20 & 4 \\ 16 & 8 \end{bmatrix}$$

矩阵的乘法不满足交换律，即 $AB \neq BA$，但是满足结合律，也就是说，$(AB)C = A(BC)$。

2.3.2 变换矩阵

Unity 中常用的变换矩阵包括平移矩阵、缩放矩阵和旋转矩阵，用于在三维空间中对物体进行基本的变换操作，它们都是 4 阶矩阵。

1. 平移矩阵和缩放矩阵

平移矩阵用于改变物体在三维空间中的位置，通过将平移向量 (T_x, T_y, T_z) 添加到平移矩阵的最后一列，可以实现物体在 x、y、z 轴上的移动。当平移矩阵与表示物体当前位置的矩阵相乘时，得到的结果矩阵将表示物体平移后的新位置。

设点 P 的坐标为 (x, y, z)，将其平移到点 (x', y', z')，在 x、y 和 z 轴上平移的距离分别为 T_x、T_y、T_z，那么 (x', y', z') 的坐标等于 $(x+T_x, y+T_y, z+T_z)$。

如果平移矩阵为

$$\begin{bmatrix} a & b & c \\ d & e & f \\ g & h & i \end{bmatrix}$$

那么平移过程表示为

$$\begin{bmatrix} x' \\ y' \\ z' \end{bmatrix} = \begin{bmatrix} a & b & c \\ d & e & f \\ g & h & i \end{bmatrix} * \begin{bmatrix} x \\ y \\ z \end{bmatrix}$$

根据矩阵乘法可以得出 $x' = ax + by + cz$，可以看出 3×3 的方阵无法表示平移，因此为它们扩充一个维度，以齐次坐标形式表示，以便加入常量 T_x、T_y 和 T_z，即点 P 为 $(x, y, z, 1)$，平移后的坐标点为 $(x', y', z', 1)$，平移矩阵设为

$$\begin{bmatrix} a & b & c & d \\ e & f & g & h \\ i & j & k & l \\ m & n & o & p \end{bmatrix}$$

平移过程表示为

$$\begin{bmatrix} x' \\ y' \\ z' \\ 1 \end{bmatrix} = \begin{bmatrix} a & b & c & d \\ e & f & g & h \\ i & j & k & l \\ m & n & o & p \end{bmatrix} * \begin{bmatrix} x \\ y \\ z \\ 1 \end{bmatrix}$$

根据矩阵乘法得知 $x' = ax + by + cz + d = x + T_x$，$y' = ex + fy + gz + h = y + T_y$，$z' = ix + jy + kz + l = z + T_z$，$1 = mx + ny + oz + p$，由此可以推导出 $a=1, b=0, c=0, d=T_x, e=0, f=1, g=0, h=T_y, i=0, j=0, k=1, l=T_z, m=0, n=0, o=0, p=1$，因此平移矩阵为

$$\begin{bmatrix} 1 & 0 & 0 & T_x \\ 0 & 1 & 0 & T_y \\ 0 & 0 & 1 & T_z \\ 0 & 0 & 0 & 1 \end{bmatrix}$$

同理可推导出缩放矩阵为

$$\begin{bmatrix} S_x & 0 & 0 & 0 \\ 0 & S_y & 0 & 0 \\ 0 & 0 & S_z & 0 \\ 0 & 0 & 0 & 1 \end{bmatrix}$$

当缩放矩阵与表示物体当前大小的矩阵相乘时，得到的结果矩阵将表示物体缩放后的新大小。如果缩放系数 $S_x = S_y = S_z$，则称这样的缩放为统一缩放，否则称为非统一缩放。从外观上来看，统一缩放是扩大或缩小整个物体，而非统一缩放会拉伸或挤压物体。

2. 旋转矩阵

旋转矩阵用于改变物体在三维空间中的方向，当旋转矩阵与表示物体当前方向的矩阵相乘时，得到的结果矩阵将表示物体旋转后的新方向。

旋转矩阵在 3 种常见的变换矩阵中是最复杂的，在了解三维的旋转矩阵前，先来看二维

的旋转矩阵。

设点 P 坐标为 (x, y)，将其沿从 x 轴到 y 轴的方向，即逆时针，旋转 β 角度，旋转后的坐标为 (x', y')，如图 2-6 所示。

旋转前和旋转后的向量长度是不变的，设长度为 m，根据三角函数定理可以得知：

$$x = m\cos\alpha, y = m\sin\alpha$$
$$x' = m\cos(\alpha + \beta), y' = m\sin(\alpha + \beta)$$

根据正弦、余弦的两角和公式，如下所示。

$$\sin(\alpha + \beta) = \sin\alpha\cos\beta + \cos\alpha\sin\beta \quad (2\text{-}12)$$
$$\cos(\alpha + \beta) = \cos\alpha\cos\beta - \sin\alpha\sin\beta \quad (2\text{-}13)$$

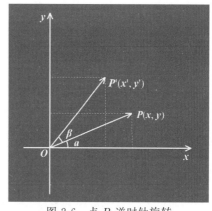

图 2-6　点 P 逆时针旋转

可以得知：

$$x' = m(\cos\alpha\cos\beta - \sin\alpha\sin\beta) = m\cos\alpha\cos\beta - m\sin\alpha\sin\beta$$
$$y' = m(\sin\alpha\cos\beta + \cos\alpha\sin\beta) = m\sin\alpha\cos\beta + m\cos\alpha\sin\beta$$

又由于 $x = m\cos\alpha, y = m\sin\alpha$，所以

$$x' = x\cos\beta - y\sin\beta$$
$$y' = x\sin\beta + y\cos\beta$$

根据矩阵的乘法，可以得知：

$$\begin{bmatrix} x' \\ y' \end{bmatrix} = \begin{bmatrix} \cos\beta & -\sin\beta \\ \sin\beta & \cos\beta \end{bmatrix} \begin{bmatrix} x \\ y \end{bmatrix}$$

因此，二维的旋转矩阵就是：

$$\begin{bmatrix} \cos\beta & -\sin\beta \\ \sin\beta & \cos\beta \end{bmatrix}$$

有了二维的旋转矩阵便可以推导出三维的旋转矩阵，在此之前，首先要了解，虽然 Unity 中默认坐标系是左手坐标系，但是在处理旋转时实际上遵循的是右手法则，这意味着如果将一个向量沿着 z 轴旋转 90°，则应该使用右手法则来判断旋转后的向量位置。

对于单位矩阵，绕哪个坐标轴旋转，则该坐标轴对应的元素保持不变，根据这一规律，将二维旋转矩阵中的元素填入相应的位置即可得到三维旋转矩阵。

绕 z 轴逆时针旋转 β 角度，z 保持不变，原来的变换与二维 x 向 y 旋转 β 角度等价，所以将二维旋转矩阵直接应用在坐标的 x、y 分量上：

$$\begin{bmatrix} x' \\ y' \\ z' \\ 1 \end{bmatrix} = \begin{bmatrix} \cos\beta & -\sin\beta & 0 & 0 \\ \sin\beta & \cos\beta & 0 & 0 \\ 0 & 0 & 1 & 0 \\ 0 & 0 & 0 & 1 \end{bmatrix} * \begin{bmatrix} x \\ y \\ z \\ 1 \end{bmatrix}$$

同理，绕 x 轴逆时针旋转 β 角度，将二维旋转矩阵直接应用在 y、z 分量上：

$$\begin{bmatrix} x' \\ y' \\ z' \\ 1 \end{bmatrix} = \begin{bmatrix} 1 & 0 & 0 & 0 \\ 0 & \cos\beta & -\sin\beta & 0 \\ 0 & \sin\beta & \cos\beta & 0 \\ 0 & 0 & 0 & 1 \end{bmatrix} * \begin{bmatrix} x \\ y \\ z \\ 1 \end{bmatrix}$$

绕 y 轴逆时针旋转 β 角度，y 保持不变，原来的变换与二维 z 向 x 旋转 β 角度方向一致，此时二维旋转矩阵作用在 z、x 分量上，所以需要将其稍做处理，将 $-\beta$ 代入，使其作用在 x、z 分量上：

$$\begin{bmatrix} x' \\ y' \\ z' \\ 1 \end{bmatrix} = \begin{bmatrix} \cos\beta & 0 & \sin\beta & 0 \\ 0 & 1 & 0 & 0 \\ -\sin\beta & 0 & \cos\beta & 0 \\ 0 & 0 & 0 & 1 \end{bmatrix} * \begin{bmatrix} x \\ y \\ z \\ 1 \end{bmatrix}$$

第 3 章 相 机 控 制

本章介绍各类型视角控制相机脚本的实现,包括第一人称视角、第三人称视角和自由控制类型视角,自由控制类型视角又分为观察者视角和漫游视角。除此之外,本章还将介绍虚拟摄像系统 Cinemachine,并通过示例展示虚拟相机如何与 Timeline 配合使用制作镜头动画。

3.1 第一人称类型相机

FPS 类型游戏中的视角是典型的第一人称视角,它模拟人眼真实的观察效果,用户可以自由地控制镜头运动和视角变化。通常情况下,在第一人称视角中,鼠标光标会被锁定到游戏窗口的中心并隐藏,使其不可见,该功能需要通过 Cursor 类中的 visible 和 lockState 两个静态变量实现,代码如下:

```
//第 3 章/FirstPersonCameraController.cs

//初始化时隐藏鼠标光标
[SerializeField] private bool hideCursorOnStart =true;

private void Start()
{
    if (hideCursorOnStart)
    {
        Cursor.visible=false;
        Cursor.lockState=CursorLockMode.Locked;
    }
}
```

其中,visible 用于设置鼠标光标是否可见,lockState 为枚举类型,包含 3 个枚举值:None 为默认状态;Locked 表示光标被锁定在游戏窗口的中心;Confined 表示光标被限制在视图中,无法移出游戏窗口。

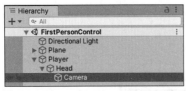

图 3-1　Player 与 Camera 的层级关系

在第一人称视角中,相机是随 Player(用户人物角色)移动的,通常情况下会将相机设为 Player 的子物体,如图 3-1 所示,无须在代码中控制其坐标值,只需通过代码控制相机的旋转值,使用户实现镜头操控自由。

实时获取鼠标在水平、垂直方向上的输入,视角跟随鼠标移动方向进行转动。实时获取输入的代码可以写在 MonoBehaviour 的生命周期函数 Update()中,需要注意的是,该函数每帧被调用一次,而由于不同时间或不同设备的性能不同,该函数每秒被调用的次数也会不同,因此为了保证匀速变化,需要将获取的输入值乘以 Time 类中的静态变量 deltaTime,该变量表示上一帧到当前帧的时间间隔,以秒为单位,Update()执行的次数越少,该时间间隔对应的值也就越大。除此之外,为了灵活地控制移动的灵敏度,通常还会乘以表示灵敏度的变量,代码如下:

```csharp
//第 3 章/FirstPersonCameraController.cs

//灵敏度
[SerializeField, Range(1f, 10f)]
private float sensitivity=3f;

private void Update()
{
    float horizontal=Input.GetAxis("Mouse X")
        * Time.deltaTime * 100f * sensitivity;
    float vertical=Input.GetAxis("Mouse Y")
        * Time.deltaTime * 100f * sensitivity;
}
```

相机在水平方向上的角度不受任何限制,而在垂直方向上的角度需要有最大值、最小值的限制,防止出现镜头翻转的现象,如图 3-2 所示。

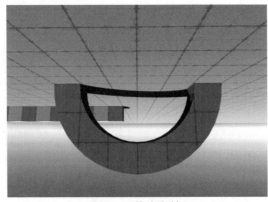

图 3-2　镜头翻转

使用两个 float 类型变量 rotY、rotX 处理鼠标在水平方向和垂直方向上的角度值，形成的欧拉角(rotX,rotY,0f)便是相机的目标角度值，最终通过 Quaternion 类中的静态函数 Euler()将该欧拉角转换为四元数，代码如下：

```
//第 3 章/FirstPersonCameraController.cs

//垂直方向角度最小值限制
[SerializeField, Range(-80f,-10f)]
private float rotXMinLimit=-40f;
//垂直方向角度最大值限制
[SerializeField, Range(10f, 80f)]
private float rotXMaxLimit=70f;
private float rotX, rotY;

private void Update()
{
    //...
    rotY+=horizontal;
    rotX-=vertical;
    rotX=Mathf.Clamp(rotX, rotXMinLimit, rotXMaxLimit);
    Quaternion targetRotation=Quaternion.Euler(rotX, rotY, 0f);
    transform.rotation=targetRotation;
}
```

3.2 第三人称类型相机

在第三人称视角中，相机跟随人物角色进行移动，人物角色的驱动控制在 Update()函数中实现，而相机跟随的逻辑通常写在 LateUpdate()函数中，该函数晚于 Update()函数执行，这样可以确保相机的运动始终在人物运动之后执行。

相机在水平方向上的朝向可以与角色的朝向保持一致，如图 3-3(a)所示，也可以由用户自由控制朝向，观察角色的不同方位，如图 3-3(b)所示，下面介绍这两种视角控制的实现方式。

3.2.1 通过角色朝向控制视角

如果想要让相机在水平方向上的朝向与角色朝向保持一致，则不需要获取鼠标在水平方向上的输入，只需根据人物角色的欧拉角设置相机的欧拉角。如果允许相机在垂直方向上进行一定的旋转，与第一人称类型相机一致，则需要对最大值和最小值进行限制，代码如下：

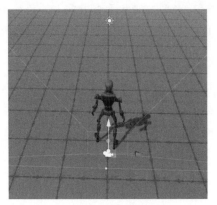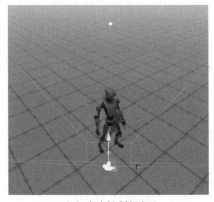

（a）朝向人物角色的前方　　　　　　　　（b）自由控制朝向

图 3-3　相机的朝向

```
//第 3 章/ ThirdPersonCameraController.cs

//垂直方向角度最小值限制
[SerializeField, Range(-80f,-10f)]
private float rotXMinLimit=-40f;
//垂直方向角度最大值限制
[SerializeField, Range(10f, 80f)]
private float rotXMaxLimit=70f;
private float rotX;

private void LateUpdate()
{
    //鼠标右键按下时旋转视角
    if (Input.GetMouseButton(1))
    {
        vertical=Input.GetAxis("Mouse Y")
            * Time.deltaTime * 100f * sensitivity;
        rotX-=vertical;
        rotX=Mathf.Clamp(rotX, rotXMinLimit, rotXMaxLimit);
    }
    Quaternion targetRotation=Quaternion.Euler(
        rotX, avatar.eulerAngles.y, 0f); //y 值取人物角色的欧拉角 y 值
}
```

为旋转添加平滑过渡的过程，使用 Quaternion 类中的插值函数 Lerp()，根据旋转速度向目标值进行插值运算，代码如下：

```
targetRotation=Quaternion.Lerp(transform.rotation,
```

```
    targetRotation, Time.deltaTime * rotationSpeed);
```

假设通过鼠标滚轮可以调整相机与人物角色之间的距离,那么需要获取鼠标滚轮的滚动值,并动态地调整控制距离的变量,但是距离不能无限制地拉近或拉远,同样应该由最大值和最小值进行限制。

通常情况下,鼠标滚轮向上滚动时,相机向前方移动,鼠标滚轮向下滚动时,相机向后方移动,也可以通过一个bool类型的变量,控制是否反转移动的方向,当其为true时,为输入值乘以-1,代码如下:

```
//第3章/ ThirdPersonCameraController.cs

//鼠标滚轮灵敏度
[SerializeField]
private float scollSensitivity=2f;
//是否反转滚动方向
[SerializeField]
private bool invertScrollDirection;
//最小距离
[SerializeField]
private float minDistanceLimit=2f;
//最大距离
[SerializeField]
private float maxDistanceLimit=5f;
//当前距离
private float currentDistance=2f;
//目标距离
private float targetDistance;

private void Start()
{
    currentDistance=Mathf.Clamp(currentDistance,
        minDistanceLimit, maxDistanceLimit);
    targetDistance=currentDistance;
}
private void LateUpdate()
{
    //...
    //鼠标滚轮滚动时改变距离
    currentDistance-=Input.GetAxis("Mouse ScrollWheel")
        * Time.deltaTime * 100f * scollSensitivity
        * (invertScrollDirection ?-1f : 1f);
```

```
    //距离限制
    currentDistance=Mathf.Clamp(currentDistance,
        minDistanceLimit, maxDistanceLimit);
    //插值
    targetDistance=Mathf.Lerp(targetDistance,
        currentDistance, Time.deltaTime);
}
```

相机的位置根据目标旋转值向后方平移目标距离,并加上人物角色的坐标值得到。由于人物角色的轴心点在底部,因此再加入一个控制高度的变量,代码如下:

```
//第 3 章 / ThirdPersonCameraController.cs

//高度
[SerializeField, Range(1f, 5f)]
private float height=2f;

private void LateUpdate()
{
    //...
    //目标位置
    Vector3 targetPosition=targetRotation * Vector3.back
        * targetDistance +avatar.position +Vector3.up * height;
    //赋值
    transform.rotation=targetRotation;
    transform.position=targetPosition;
}
```

3.2.2　通过用户输入控制视角

在 3.2.1 节中实现了通过角色朝向控制的第三人称视角,如果想要通过用户输入控制视角,自由地观察人物角色的各方位,则需要在此基础上获取鼠标在水平方向上的输入,将输入值应用于欧拉角 y 值。增加一个 bool 类型变量,用于表示是否由角色朝向控制相机水平方向上的朝向,当其为 true 时,使用人物角色的欧拉角 y 值;当其为 false 时,根据输入值设置欧拉角 y 值,代码如下:

```
//第 3 章 / ThirdPersonCameraController.cs

//...
private float rotY;
[SerializeField] private bool backOfAvatar;
```

```
private void LateUpdate()
{
    //当鼠标右键按下时旋转视角
    if (Input.GetMouseButton(1))
    {
        float horizontal=Input.GetAxis("Mouse X")
            * Time.deltaTime * 100f * sensitivity;
        float vertical=Input.GetAxis("Mouse Y")
            * Time.deltaTime * 100f * sensitivity;
        rotY+=horizontal;
        rotX-=vertical;
        rotX=Mathf.Clamp(rotX, rotXMinLimit, rotXMaxLimit);
    }
    Quaternion targetRotation=Quaternion.Euler(
        rotX, backOfAvatar ? avatar.eulerAngles.y : rotY, 0f);
    //...
}
```

接下来实现这样一个功能,当相机的视角根据用户输入进行旋转时,人物角色的头部同步进行转动,如图3-4所示。

头部转动的角度需要进行限制,避免出现头部转动到身后等怪异现象,如图3-5所示。当角度达到限制值时,可以自动回正头部朝向,不再跟随相机视角。

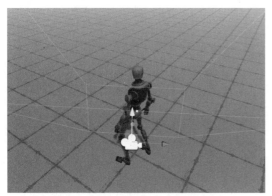

图3-4 角色头部随相机视角转动

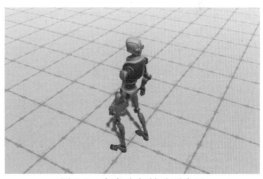

图3-5 角色头部转动到身后

相机坐标加相机前方一定单位获得的点是头部需要看向的点,该点减去角色头部坐标点形成看向的方向,根据该方向可以求得目标角度值,判断角度值是否在限制值范围内。

由于旋转是360°的,以180°为例,当角度为-180°时与朝向和角度为180°时是一样的,因此需要一种方法进行调整。当角度大于180°时,将其减去360°,当角度小于-180°时,将其加上360°,这样使角度的取值范围始终为[-180,180],代码如下:

```csharp
//第 3 章/HeadTracker.cs
private Camera mainCamera; //主相机
private Transform head; //头部
[SerializeField] private Animator animator;
[Tooltip("水平方向上的角度限制"), SerializeField]
private Vector2 horizontalAngleLimit=new Vector2(-70f, 70f);
[Tooltip("垂直方向上的角度限制"), SerializeField]
private Vector2 verticalAngleLimit=new Vector2(-60f, 60f);

private void Start()
{
    mainCamera=Camera.main !=null
        ? Camera.main : FindObjectOfType<Camera>();
    head=animator.GetBoneTransform(HumanBodyBones.Head);
}
//获取看向的位置
private Vector3 GetLookAtPosition()
{
    //相机前方一定单位的位置
    Vector3 position=mainCamera.transform.position
        +mainCamera.transform.forward * 100f;
    //看向的方向
    Quaternion lookRotation=Quaternion.LookRotation(
        position-head.position, animator.transform.up);
    Vector3 angle=lookRotation.eulerAngles
        -animator.transform.eulerAngles;
    float x=NormalizeAngle(angle.x);
    float y=NormalizeAngle(angle.y);
    //是否在限制值范围内
    bool isInRange=x>=verticalAngleLimit.x && x<=verticalAngleLimit.y
        && y>=horizontalAngleLimit.x && y<=horizontalAngleLimit.y;
    return isInRange ? position
        : (head.position +animator.transform.forward);
}
//角度标准化
private float NormalizeAngle(float angle)
{
    if (angle>180) angle-=360f;
    else if (angle<-180) angle+=360f;
    return angle;
}
```

获得看向的位置后,在 LateUpdate()函数中让头部看向该位置,代码如下:

```csharp
//第 3 章/HeadTracker.cs

private float angleX;
private float angleY;
[Tooltip("插值速度"), SerializeField]
private float lerpSpeed=5f;

private void LateUpdate()
{
    LookAtPosition(GetLookAtPosition());
}
//<summary>
//看向某点
//</summary>
//<param name="position">看向的点</param>
public void LookAtPosition(Vector3 position)
{
    Quaternion lookRotation=Quaternion.LookRotation(
        position-head.position);
    Vector3 eulerAngles=lookRotation.eulerAngles
        -animator.transform.rotation.eulerAngles;
    float x=NormalizeAngle(eulerAngles.x);
    float y=NormalizeAngle(eulerAngles.y);
    angleX=Mathf.Clamp(Mathf.Lerp(angleX, x,
        Time.deltaTime * lerpSpeed),
        verticalAngleLimit.x, verticalAngleLimit.y);
    angleY=Mathf.Clamp(Mathf.Lerp(angleY, y,
        Time.deltaTime * lerpSpeed),
        horizontalAngleLimit.x, horizontalAngleLimit.y);
    Quaternion rotY=Quaternion.AngleAxis(
        angleY, head.InverseTransformDirection(animator.transform.up));
    head.rotation *=rotY;
    Quaternion rotX=Quaternion.AngleAxis(
        angleX, head.InverseTransformDirection(
            animator.transform.TransformDirection(Vector3.right)));
    head.rotation *=rotX;
}
```

3.3 自由控制类型相机

本节介绍两种自由类型相机控制脚本的实现，分别为观察者视角控制和漫游视角控制，二者的核心区别在于，前者在全局坐标系中进行前、后、左、右及上、下方向的移动，视角围绕与地面的交点旋转，后者在本地坐标系中沿自身的前、后、左、右及上、下方向进行移动，视角围绕自身进行旋转，它们适用于不同的场景。

3.3.1 观察者视角控制

在观察者视角控制中，可以通过键盘上的 W、A、S、D、Q 和 E 键进行移动，对应方向分别为前、左、后、右、下和上。在方法中声明一个 Vector3 类型的变量，表示移动的方向，当对应的按键被按下时，为该变量赋值，代码如下：

```
//第 3 章/GodlikeCameraController.cs

//移动方向
Vector3 motion=Vector3.zero;
if (Input.GetKey(KeyCode.W)) //向前移动
    motion+=Vector3.forward;
if (Input.GetKey(KeyCode.S)) //向后移动
    motion+=Vector3.back;
if (Input.GetKey(KeyCode.A)) //向左移动
    motion+=Vector3.left;
if (Input.GetKey(KeyCode.D)) //向右移动
    motion+=Vector3.right;
if (Input.GetKey(KeyCode.Q)) //向下移动
    motion+=Vector3.down;
if (Input.GetKey(KeyCode.E)) //向上移动
motion+=Vector3.up;
```

还可以通过按住鼠标左键进行拖曳，根据鼠标拖曳产生的偏移量进行移动。声明一个全局的 Vector2 类型变量，用于记录上一帧的鼠标位置，当前帧的鼠标位置减去上一帧的鼠标位置便可以获得鼠标移动的偏移量，代码如下：

```
//第 3 章/GodlikeCameraController.cs

//用于记录上一帧的鼠标位置
private Vector2 lastMousePosition;

private void Update()
```

```
{
    //...
    //鼠标左键是否被按住,鼠标位置是否处于窗口内
    if (Input.GetMouseButton(0)
        && Screen.safeArea.Contains(Input.mousePosition))
    {
        //当前帧的鼠标位置减去上一帧的鼠标位置得到偏移量
        Vector3 mousePosDelta=new Vector2(
            Input.mousePosition.x-lastMousePosition.x,
            Input.mousePosition.y-lastMousePosition.y);
        motion+=Vector3.left * mousePosDelta.x
            +Vector3.back * mousePosDelta.y;
    }
    //记录鼠标位置
    lastMousePosition=Input.mousePosition;
}
```

除此之外,还可以通过鼠标滚轮向视角的前方或后方进行移动,获取移动方向的代码如下:

```
float wheelValue=Input.GetAxis("Mouse ScrollWheel");
motion+=wheelValue==0 ? Vector3.zero
    : ((wheelValue>0 ? Vector3.forward : Vector3.back)
        * (invertScrollDirection ? -1f : 1f));
```

获取移动的方向后,计算目标坐标值,由于通过鼠标滚轮移动是沿自身的前方或后方进行移动的,而其他移动是沿世界空间中的前、后、左、右等方向进行移动的,因此需要进行相应区分。

假设通过键盘左侧的 Shift 按键可以加速移动,那么在将移动方向归一化后,可以通过向量数乘得到最终的移动向量。通过旋转四元数乘以移动向量得到目标坐标值,从当前坐标值向目标坐标值进行插值运算,实现平滑移动,代码如下:

```
//移动速度
[SerializeField]
private float moveSpeed=50f;
//加速系数,当 Shift 键被按下时起作用
[SerializeField]
private float boostFactor=3.5f;
//目标坐标值
private Vector3 targetPos;
//目标角度值
```

```csharp
private Vector3 targetRot;
//插值速度
[SerializeField]
private float lerpSpeed=10f;

private void OnEnable()
{
    targetPos=transform.position;
    targetRot=transform.eulerAngles;
    lastMousePosition=Input.mousePosition;
}
private void Update()
{
    //...
    //归一化
    motion=motion.normalized;
    //按住左 Shift 键时移动加速
    if (Input.GetKey(KeyCode.LeftShift))
        motion * =boostFactor;
    motion * =Time.deltaTime * moveSpeed;
    targetPos+=(wheelValue !=0f ? Quaternion.Euler(targetRot)
        : Quaternion.Euler(0f, targetRot.y, targetRot.z)) * motion;
    transform.position=Vector3.Lerp(transform.position,
        targetPos, Time.deltaTime * lerpSpeed);
    transform.rotation=Quaternion.Lerp(transform.rotation,
        Quaternion.Euler(targetRot), Time.deltaTime * lerpSpeed);
    //记录鼠标位置
    lastMousePosition=Input.mousePosition;
}
```

当按住鼠标右键拖曳鼠标时，视角绕与地平面的交点进行旋转，交点在鼠标右键开始被按下的帧中进行计算，计算过程需要用到向量的点乘运算。

2.2.4 节介绍了向量的点乘运算，如果两个向量均是单位化的向量，模长均为 1，则点乘结果就是这两个向量夹角的余弦值。基于此，可以通过 transform.forward 与 Vector3.down 向量求得相机前方与世界空间中正下方向量的余弦值，从而可以根据该余弦值和相机的坐标 y 值求得从相机位置沿相机前方到地平面的距离，最终得到交点坐标，代码如下：

```csharp
//第 3 章/GodlikeCameraController.cs

//鼠标右键拖动时围绕该点进行旋转
private Vector3 center=Vector3.zero;
```

```
private void Update()
{
    //鼠标右键开始被按下时计算视角围绕的点
    if (Input.GetMouseButtonDown(1))
    {
        //相机前方与世界空间中正下方向量的余弦值
        float cos=Vector3.Dot(transform.forward, Vector3.down);
        //根据余弦值和相机坐标 y 值求得从相机位置沿相机前方到地平面的距离
        float distance=transform.position.y / cos;
        distance=distance<0f ? 0f : distance;
        center=transform.position +transform.forward * distance;
        center=!float.IsNaN(center.magnitude) ? center : Vector3.zero;
    }
    //...
}
```

运行结果如图 3-6 所示。

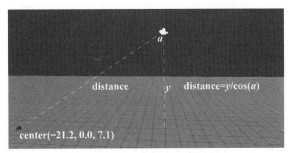

图 3-6　求解中心点

当鼠标右键持续被按下时，调用 Transform 类中的 RotateAround()方法即可实现按交点进行旋转，代码如下：

```
//第 3 章/GodlikeCameraController.cs

private void Update()
{
    //...
    else if (Input.GetMouseButton(1))
    {
        //当前帧与上一帧鼠标位置发生的偏移量
        Vector3 mousePosDelta=new Vector2(
            Input.mousePosition.x-lastMousePosition.x,
```

```
            Input.mousePosition.y-lastMousePosition.y);
        //鼠标右键被按下拖动时绕交点旋转
        transform.RotateAround(center, Vector3.up, mousePosDelta.x
            * Time.deltaTime * mouseMovementSensitivity);
        transform.RotateAround(center, transform.right,-mousePosDelta.y
            * Time.deltaTime * mouseMovementSensitivity);
        targetPos=transform.position;
        targetRot=transform.eulerAngles;
    }
    //...
}
```

在 RTS 类型的游戏中通常会有这样的功能,当鼠标光标移动到屏幕边缘时,视角也会向该边缘方向进行移动,实现该功能首先要定义屏幕的边缘,假设边缘宽度为 2,屏幕分辨率为 1920×1080,那么鼠标坐标 x 值在[0,2]或[1918,1920]取值范围内时,或者坐标 y 值在[0,2]或[1078,1080]取值范围内时,可以判定鼠标光标在屏幕边缘,代码如下:

```
//第3章/GodlikeCameraController.cs

//是否启用鼠标光标处于屏幕边缘时向外移动
[SerializeField]
private bool enableScreenEdgeMove;
//该值用于定义屏幕边缘区域
[SerializeField]
private float screenEdgeDefine=2f;

private void Update()
{
    //...
    if (enableScreenEdgeMove)
    {
        bool isMouseOnHorizontalScreenEdge =
            Input.mousePosition.x<=screenEdgeDefine ||
            Input.mousePosition.x>=Screen.width-screenEdgeDefine;
        bool isMouseOnVerticalScreenEdge =
            Input.mousePosition.y<=screenEdgeDefine ||
            Input.mousePosition.y>=Screen.height-screenEdgeDefine;
        if (isMouseOnHorizontalScreenEdge)
            motion+=Input.mousePosition.x<=screenEdgeDefine
                ?Vector3.left : Vector3.right;
        if (isMouseOnVerticalScreenEdge)
```

```
            motion+=Input.mousePosition.y<=screenEdgeDefine
                ?Vector3.back : Vector3.forward;
    }
    //...
}
```

3.3.2 漫游视角控制

漫游视角控制与观察者视角控制的实现过程类似,只不过在漫游视角控制中处理移动时是沿相机的自身方向进行移动,这与观察者视角控制中通过鼠标滚轮移动的方式一致。处理旋转时以自身为轴心进行旋转,不需要求鼠标位置发生的偏移量及中心点等,根据鼠标在水平方向和垂直方向上的输入值进行旋转即可,代码如下:

```
//第 3 章/RoamCameraController.cs

using UnityEngine;

//<summary>
//漫游类型相机控制
//</summary>
public class RoamCameraController : MonoBehaviour
{
    //移动速度
    [SerializeField]
    private float moveSpeed=50f;
    //加速系数,当 Shift 键被按下时起作用
    [SerializeField]
    private float boostFactor=3.5f;
    //是否反转方向,用于鼠标滚轮移动
    [SerializeField]
    private bool invertScrollDirection=false;
    //鼠标转动的灵敏度
    [Range(0.1f, 20f), SerializeField]
    private float mouseMovementSensitivity=10f;
    //反转水平转动方向
    [SerializeField] private bool invertY=false;
    //插值速度
    [SerializeField]
    private float lerpSpeed=10f;
    //目标坐标值
    private Vector3 targetPos;
```

```csharp
//目标旋转值
private Vector3 targetRot;

private void OnEnable()
{
    targetPos=transform.position;
    targetRot=transform.eulerAngles;
}
private void Update()
{
    //当按住鼠标右键拖动时旋转视角
    if (Input.GetMouseButton(1))
    {
        float x=Input.GetAxis("Mouse X");
        float y=Input.GetAxis("Mouse Y");
        targetRot+=new Vector3(y * (invertY ? 1f :-1f), x, 0f)
            * mouseMovementSensitivity * Time.deltaTime * 100f;
    }
    //移动方向
    Vector3 motion=Vector3.zero;
    if (Input.GetKey(KeyCode.W)) //向前移动
        motion+=Vector3.forward;
    if (Input.GetKey(KeyCode.S)) //向后移动
        motion+=Vector3.back;
    if (Input.GetKey(KeyCode.A)) //向左移动
        motion+=Vector3.left;
    if (Input.GetKey(KeyCode.D)) //向右移动
        motion+=Vector3.right;
    if (Input.GetKey(KeyCode.Q)) //向下移动
        motion+=Vector3.down;
    if (Input.GetKey(KeyCode.E)) //向上移动
        motion+=Vector3.up;

    float wheelValue=Input.GetAxis("Mouse ScrollWheel");
    motion+=wheelValue==0 ? Vector3.zero
        : (wheelValue>0 ? Vector3.forward : Vector3.back)
            * (invertScrollDirection ? -1 : 1);

    motion=motion.normalized;
    if (Input.GetKey(KeyCode.LeftShift))
        motion *=boostFactor;
```

```
            motion *=Time.deltaTime * moveSpeed;
            targetPos+=Quaternion.Euler(targetRot) * motion;
            transform.position=Vector3.Lerp(transform.position,
                targetPos, Time.deltaTime * lerpSpeed);
            transform.rotation=Quaternion.Lerp(transform.rotation,
                Quaternion.Euler(targetRot), Time.deltaTime * lerpSpeed);
        }
}
```

3.4 Cinemachine

Cinemachine 是官方在 2017 年推出的一套专门控制相机的模块工具，它解决了摄像机间的复杂控制、混合、切换等复杂数学和逻辑问题，减少了在开发过程中开发者对相机的脚本控制编写所需的时间成本，并能快速实现一些摄像机功能。该工具需要在 Package Manager 中下载导入，如图 3-7 所示。

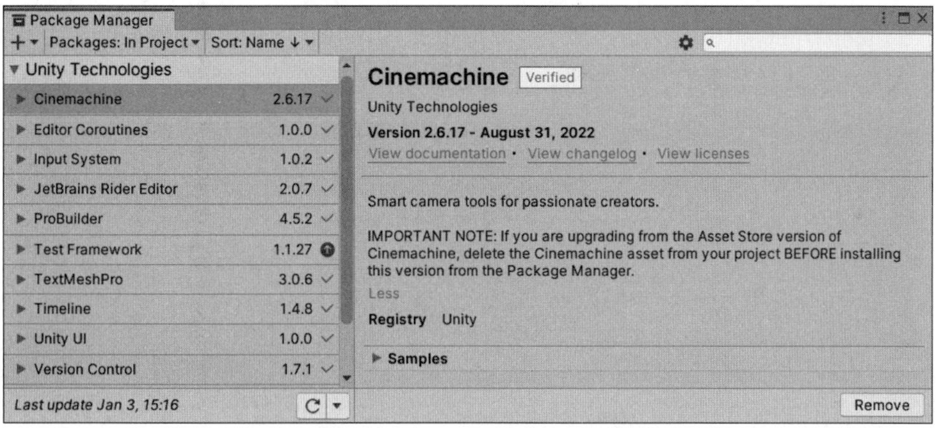

图 3-7　Cinemachine Package

在 Cinemachine 中，虚拟相机（Virtual Camera）是一个重要的概念。创建虚拟相机后，场景中的原相机会自动添加一个 CinemachineBrain 组件，如图 3-8 所示，该组件是整个模块的核心组件，可以称为"大脑"，它挂载于 Camera 组件所在的对象上，监控着场景中所有为活跃状态的虚拟相机，以实现镜头的切换及控制。此外，开发者还可以使用 Timeline 来控制虚拟相机，实现效果丰富的镜头动画。

3.4.1　基于虚拟相机实现第三人称视角

Cinemachine 中提供了第三人称相机解决方案，导入 Cinemachine 工具后，可以发现 Unity 窗口顶部多了一个 Cinemachine 菜单栏，通过菜单栏中的 Create Virtual Camera 选

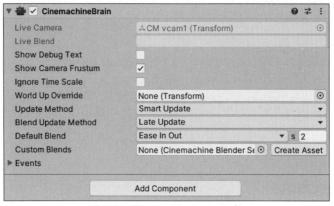

图 3-8　CinemachineBrain 组件

项即可在场景中创建一个虚拟相机。

创建完成后,可以发现场景发生了两个变化。第一,在主相机的层级中,出现了一个图标,如图 3-9 所示,该图标表示主相机挂载了 CinemachineBrain 组件,一个场景中只存在一个 CinemachineBrain 组件。第二,场景中多了一个新的游戏物体 CM vcam1,该游戏物体上挂载了 CinemachineVirtualCamera 组件,即虚拟相机,如图 3-10 所示,如果继续创建,则场景中便会多出新的挂载虚拟相机组件的游戏物体 CM vcam2,以此类推。

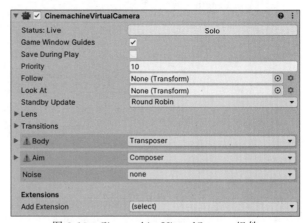

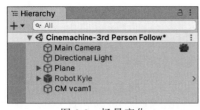

图 3-9　场景变化　　　　　图 3-10　CinemachineVirtualCamera 组件

如果要将虚拟相机作为第三人称视角相机使用,则需要为 Follow 属性赋值,如图 3-11

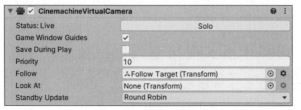

图 3-11　虚拟相机跟随的目标

所示，它表示相机跟随的目标。

设置跟随的目标后，将 Body 的类型设置为 3rd Person Follow，如图 3-12 所示，它表示将虚拟相机作为第三人称视角跟随相机使用，Body 中的属性详解见表 3-1。

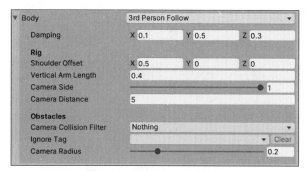

图 3-12　3rd Person Follow

表 3-1　3rd Person Follow Body 属性详解

属　　性	详　　解
Damping	相机跟随的阻尼值，值越大，相机到达目标位置的速度越慢
Shoulder Offset	肩部偏移值（相对于跟随目标位置的偏移量）
Vertical Arm Length	手相对于肩部的垂直偏移值，当相机在垂直方向上旋转时，该值会影响跟随目标在屏幕中的位置
Camera Side	指定相机偏向哪个肩部，取值范围为 0～1，0 表示左侧，1 表示右侧
Camera Distance	相机与跟随目标的距离
Camera Collision Filter	障碍物检测层级（相机将避开通过这些层级设定的障碍物）
Ignore Tag	设置为此标签的障碍物将被忽略
Camera Radius	相机半径（用于指定相机离障碍物的距离）

当跟随目标是角色本身时，相机朝向与角色朝向保持一致，假设想要实现通过用户输入使相机围绕角色旋转，可以将跟随目标设置为独立于角色旋转的空物体，通过旋转该空物体实现相机围绕角色进行旋转。

3.4.2　轨道路径与推轨相机

通过 Cinemachine 工具可以实现复杂的镜头动画，如图 3-13 所示，将 CinemachineVirtualCamera 组件中 Body 的类型设置为 Tracked Dolly 类型，可以实现让虚拟相机沿自定义的轨道路径移动，该类型 Body 中的属性详解见表 3-2。

启用 Auto Dolly 后，虚拟相机的位置会自动动画化到路径上最接近跟随目标的位置，这也意味着，使用 Tracked Dolly 时，不仅需要提供一个轨道路径，还需要指定一个跟随的目标。

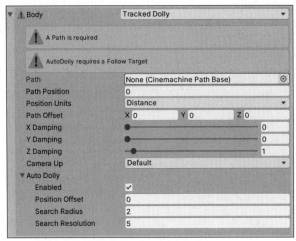

图 3-13 Tracked Dolly

表 3-2 Tracked Dolly Body 属性详解

属　　性	详　　解
Path	轨道路径
Path Position	虚拟相机在路径中的位置
Position Units	路径位置的度量单位
Path Offset	虚拟相机相对于路径的位置
X Damping	虚拟相机到达目标位置在 x 轴上的阻尼值
Y Damping	虚拟相机到达目标位置在 y 轴上的阻尼值
Z Damping	虚拟相机到达目标位置在 z 轴上的阻尼值
Camera Up	设置虚拟相机向上方向的类型
Auto Dolly	自动推轨的相关设置

　　工具中提供了两种轨道路径组件，分别为 CinemachinePath 和 CinemachineSmoothPath。路径通过贝塞尔曲线实现，如果切线设置不当，则会影响路径动画的平滑性和连续性，两种类型路径组件的主要区别在于 CinemachineSmoothPath 会自动设置切线，以确保完全平滑。

　　创建一个新的游戏物体并为其添加 CinemachineSmoothPath 组件，在该组件的检视面板中编辑路径点，如图 3-14 所示，也可以在场景中进行编辑，该组件对应的编辑器类中通过创建坐标操控柄为在场景中编辑路径点提供了支持。

　　路径点编辑完成后，当选中组件所在的游戏物体时，路径以轨道的形状展示在场景中，如图 3-15 所示。

　　创建另一个游戏物体并为其添加 CinemachineDollyCart 组件，将其作为虚拟相机的跟

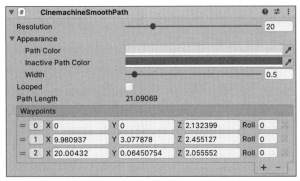

图 3-14　CinemachineSmoothPath

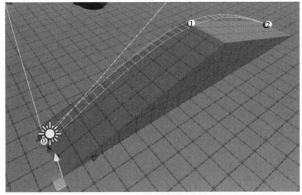

图 3-15　轨道路径

随目标,该组件用于实现沿轨道路径移动,因此需要为其设置一个轨道路径,如图 3-16 所示,可以将其理解为行驶在指定轨道上的小车。

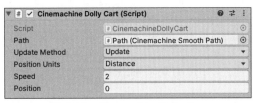

图 3-16　CinemachineDollyCart

Update Method 表示使用的刷新方法,包含 Update、FixedUpdate、LateUpdate 共 3 种类型。Position Units 与 Tracked Dolly Body 中的 Position Units 属性一致,表示路径位置的度量单位,包含 Path Units、Distance、Normalized 共 3 种类型,Path Units 表示使用路径点进行度量,例如在第 1 个路径点时,Position 为 0,在第 2 个路径点时,Position 为 1,Distance 表示使用距离进行度量,单位为 m,Normalized 表示归一化,0 表示路径开头,1 表示路径结尾。

CinemachineVirtualCamera 组件中的 Aim 属性决定了虚拟相机看向目标时使用的算

法，包含 Do nothing、Composer、Group Composer、Hard Look At、POV、Same As Follow Target 共 6 种类型。

Do nothing 表示相机不做任何旋转操作，Composer 表示将看向目标保持在相机画面中，Group Composer 的行为与 Composer 一致，但是相机将对准多个游戏对象，看向的是一个目标组，需要为 Look At 的对象挂载 CinemachineTargetGroup 组件。Hard Look At 表示将看向目标保持在相机画面的中心，POV 表示根据用户输入旋转视角，Same As Follow Target 表示将相机的旋转设置为跟随的目标的旋转，也就是与跟随目标的旋转保持一致。

如图 3-17 所示，假设虚拟相机在轨道路径上移动时始终看向场景中的某个球形游戏物体，可以将该物体作为虚拟相机 Look At 的对象，将 Aim 设置为 Composer 或 Hard Look At 类型。设置完成后运行程序，可以看到虚拟相机在轨道路径上移动的同时，视角始终看向场景中的球形物体。

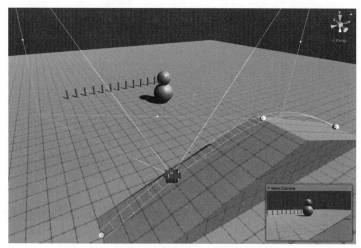

图 3-17　虚拟相机沿轨道路径移动并看向目标

3.4.3　在 Timeline 中控制镜头

在 Timeline 中可以管理多个虚拟相机的状态，通过启用、停用及混合不同的虚拟相机实现镜头之间的切换。

Timeline 中默认可以创建 Activation、Animation 等类型的轨道，在导入 Cinemachine 工具包后，便多了一个新的轨道类型，即 Cinemachine Track，如图 3-18 所示。不同类型的轨道配合使用可以实现丰富的动画效果。

1. Cinemachine Track

Cinemachine Track 需要 CinemachineBrain 组件对象，将挂载 CinemachineBrain 组件的主相机拖入赋值。

通过在时间轴中右击并单击 Add Cinemachine Shot 或者直接将挂载虚拟相机组件的游戏物体拖入即可创建新的镜头剪辑，如图 3-19 所示。

图 3-18　Create Cinemachine Track

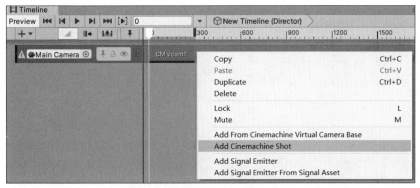

图 3-19　Add Cinemachine Shot

在示例场景中创建一条新的轨道路径,并为一个球体添加 Cinemachine Dolly Cart 组件,使其沿着路径移动,如图 3-20 所示。

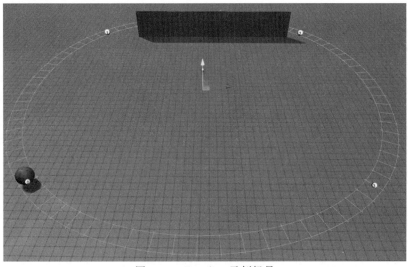

图 3-20　Timeline 示例场景

创建两个虚拟相机，将第 1 个虚拟相机的 Look At 对象设置为移动的球体，将 Aim 设为 Hard Look At 类型，用于始终看向相机。将第 2 个虚拟相机 Follow 对象设置为移动的球体，将 Body 设为 3rd Person Follow 类型，用于跟随球体移动。

将这两个虚拟相机依次拖入 Timeline 中的 Cinemachine Track 中，如图 3-21 所示。设置完成后运行程序，可以看到，首先是第 1 个虚拟相机处于活跃状态，相机根据球体运动旋转视角，始终看向球体，第 1 个镜头剪辑播放完成后，第 2 个虚拟相机进入活跃状态，相机以第三人称视角跟随球体。

当所有的镜头剪辑都播放完成后，控制权将返给 CinemachineBrain，它会选择具有最高优先级设置的虚拟相机，优先级通过 Priority 属性进行设置。

如果两个虚拟镜头之间需要平滑过渡，则需要将两个镜头剪辑的过渡部分重叠，如图 3-22 所示，斜线标识的部分即重叠区域。

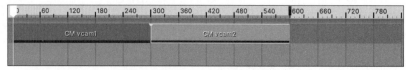

图 3-21 Cinemachine Track 设置

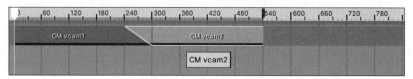

图 3-22 镜头过渡设置

2. Movement Track

值得一提的是，开发者可以像 Cinemachine 一样，添加自定义轨道类型。本节以实现自定义轨道 Movement Track 为例，介绍自定义的过程，Movement Track 类型的轨道的作用是使物体向指定的位置进行移动。

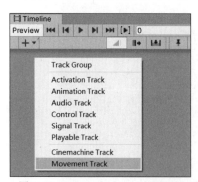

图 3-23 Create Movement Track

首先创建 Movement Track 类并继承 TrackAsset，有了这个类后，在 Timeline 中便可以创建该类型的轨道，如图 3-23 所示。

然后需要为 Movement Track 类添加 3 个特性，分别是 TrackColor、TrackBindingType 及 TrackClipType。TrackColor 用于指定轨道的颜色，TrackBindingType 用于指定绑定的对象类型，TrackClickType 用于指定轨道剪辑的类型，通过菜单 Assets/Create/Playables/Playable Asset C# Script 创建，通过这个菜单创建的类自动继承 PlayableAsset 类，代码如下：

```
//第3章/MovementPlayableAsset.cs

using UnityEngine;
using UnityEngine.Playables;

[System.Serializable]
public class MovementPlayableAsset : PlayableAsset
{
    //Factory method that generates a playable based on this asset
    public override Playable CreatePlayable(
        PlayableGraph graph, GameObject go)
    {
        return Playable.Create(graph);
    }
}
```

例如将轨道颜色设为紫色,RGB值分别为1、0、1,将绑定的对象类型设为GameObject,将轨道剪辑设为MovementPlayableAsset类型,那么Movement Track类的代码如下:

```
//第3章/MovementTrack.cs

using UnityEngine;
using UnityEngine.Timeline;

[TrackColor(1f, 0f, 1f)]
[TrackBindingType(typeof(GameObject))]
[TrackClipType(typeof(MovementPlayableAsset))]
public class MovementTrack : TrackAsset { }
```

此时创建一个Movement Track类型的轨道,可以看到轨道的颜色为紫色,并且可以将Game Object类型的对象拖入赋值,如图3-24所示。

轨道剪辑中的行为实现需要通过行为类进行定义,通过菜单Assets/Create/Playables/Playable Behaviour C# Script创建行为类,创建的类自动继承PlayableBehaviour类,并重写部分虚方法。声明移动行为所需的变量,代码如下:

图3-24 Movement Track

```
//第3章/MovementPlayableBehaviour.cs

using UnityEngine;
using UnityEngine.Playables;
```

```csharp
//A behaviour that is attached to a playable
public class MovementPlayableBehaviour : PlayableBehaviour
{
    //要移动的对象
    public Transform actor;
    //移动的目标点
    public Vector3 targetPosition;
    //起点
    private Vector3 beginPosition;

    //Called when the owning graph starts playing
    public override void OnGraphStart(
        Playable playable) { }
    //Called when the owning graph stops playing
    public override void OnGraphStop(
        Playable playable) { }
    //Called when the state of the playable is set to Play
    public override void OnBehaviourPlay(
        Playable playable, FrameData info) { }
    //Called when the state of the playable is set to Paused
    public override void OnBehaviourPause(
        Playable playable, FrameData info) { }
    //Called each frame while the state is set to Play
    public override void PrepareFrame(
        Playable playable, FrameData info) { }
}
```

在 MovementPlayableAsset 类中重写抽象方法 CreatePlayable()，用于指定轨道剪辑的行为对象，并为行为对象提供数据，代码如下：

```csharp
//第 3 章/MovementPlayableAsset.cs

using UnityEngine;
using UnityEngine.Playables;

[System.Serializable]
public class MovementPlayableAsset : PlayableAsset
{
    public ExposedReference<Transform> actor;
    public Vector3 targetPosition;
```

```
//Factory method that generates a playable based on this asset
public override Playable CreatePlayable(
    PlayableGraph graph, GameObject go)
{
    var behaviour=new MovementPlayableBehaviour()
    {
        actor=actor.Resolve(graph.GetResolver()),
        targetPosition=targetPosition
    };
    return ScriptPlayable<MovementPlayableBehaviour>
        .Create(graph, behaviour);
}
```

在 Movement Track 中创建轨道剪辑并选中,在检视面板中设置要移动的对象和目标点,如图 3-25 所示。

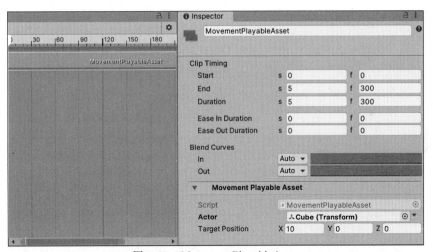

图 3-25　MovementPlayableAsset

重写行为类中的相关虚方法,在轨道剪辑播放的过程中,插值计算移动对象的位置,代码如下：

```
//第 3 章/MovementPlayableBehaviour.cs

public override void OnBehaviourPlay(Playable playable, FrameData info)
{
    //开始播放 记录起点
    if (actor !=null)
```

```
        beginPosition=actor.position;
}
public override void OnBehaviourPause(Playable playable, FrameData info)
{
    //播放结束 回到起点
    if (actor !=null)
        actor.position=beginPosition;
}
public override void ProcessFrame(
        Playable playable, FrameData info, object playerData)
{
    //播放过程插值计算当前位置
    if (actor !=null)
    {
        float duration=(float)playable.GetDuration();
        float time=(float)playable.GetTime();
        float lerpPct=time / duration;
        actor.position=Vector3.Lerp(
            beginPosition, targetPosition, lerpPct);
    }
}
```

运行程序，可以发现移动对象在轨道剪辑播放的过程中向目标位置进行移动，如图 3-26 所示。

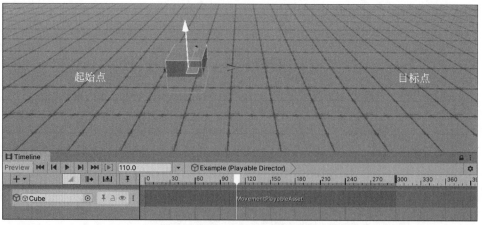

图 3-26　Movement Track 运行

第 4 章 物 理 检 测

Physics 类是用于处理物理相关功能的核心类，它提供了系统的方法，包括射线投射检测、球体投射检测和盒体投射检测等物理检测方法。本章将通过具体案例介绍常用物理检测方法的使用过程并介绍如何借助 Gizmos 实现物理检测的可视化。

4.1 射线投射检测

射线投射检测指的是 Physics 类中的 Raycast()方法，在调用该方法时，需要指定射线投射的起点及方向，或者指定表示射线的 Ray 类型变量，除此之外，还可以指定射线投射检测的最大距离及检测的层级等，因此，它有多个重载方法，代码如下：

```
public static bool Raycast(Vector3 origin, Vector3 direction);
public static bool Raycast(Vector3 origin, Vector3 direction,
    float maxDistance);
public static bool Raycast(Vector3 origin, Vector3 direction,
    float maxDistance, int layerMask);
public static bool Raycast(Vector3 origin, Vector3 direction,
    float maxDistance, int layerMask,
    QueryTriggerInteraction queryTriggerInteraction);
public static bool Raycast(Vector3 origin, Vector3 direction,
    out RaycastHit hitInfo, float maxDistance, int layerMask,
    QueryTriggerInteraction queryTriggerInteraction);
public static bool Raycast(Vector3 origin, Vector3 direction,
    out RaycastHit hitInfo, float maxDistance, int layerMask);
public static bool Raycast(Vector3 origin, Vector3 direction,
    out RaycastHit hitInfo, float maxDistance);
public static bool Raycast(Vector3 origin, Vector3 direction,
    out RaycastHit hitInfo);
public static bool Raycast(Ray ray);
public static bool Raycast(Ray ray, float maxDistance);
```

```csharp
public static bool Raycast(Ray ray, float maxDistance, int layerMask);
public static bool Raycast(Ray ray, float maxDistance, int layerMask,
    QueryTriggerInteraction queryTriggerInteraction);
public static bool Raycast(Ray ray, out RaycastHit hitInfo);
public static bool Raycast(Ray ray, out RaycastHit hitInfo,
    float maxDistance);
public static bool Raycast(Ray ray, out RaycastHit hitInfo,
    float maxDistance, int layerMask);
public static bool Raycast(Ray ray, out RaycastHit hitInfo,
    float maxDistance, int layerMask,
    QueryTriggerInteraction queryTriggerInteraction);
```

参数详解见表 4-1。

表 4-1 Raycast()方法参数详解

参数	详解
origin	射线投射的起点
direction	射线投射的方向
ray	表示射线的结构体,包含起点和方向信息
hitInfo	如果方法的返回值为 true,则表示射线投射检测到了碰撞体,hitInfo 中包含检测到的碰撞体、碰撞位置和法线方向等信息
maxDistance	用于指定射线投射检测的最大距离
layerMask	用于指定射线投射检测的层级
queryTriggerInteraction	用于指定射线投射检测是否命中触发器

4.1.1 获取鼠标单击地面位置

在游戏单位寻路功能中,通常会通过鼠标单击地面的位置获得寻路目的地,那么如何得知鼠标是否单击了地面并获取单击的位置呢?可以通过射线投射检测实现。

通过 Camera 类中的 ScreenPointToRay()方法可以从鼠标位置创建一条射线,通过这条射线进行射线投射检测,如果方法的返回值为 true,则通过 hitInfo 获取检测到的碰撞体。将地面的标签指定为 Ground,判断射线投射检测到的碰撞体的标签是否为 Ground 便可以得知鼠标是否单击了地面,代码如下:

```csharp
//第 4 章/GroundClick.cs

using UnityEngine;

public class GroundClick : MonoBehaviour
```

```csharp
{
    //主相机
    [SerializeField]
    private Camera mainCamera;

    private void Start()
    {
        if (mainCamera==null)
        {
            mainCamera=Camera.main !=null
                ? Camera.main
                : FindObjectOfType<Camera>();
        }
    }

    private void Update()
    {
        //鼠标左键单击
        if (Input.GetMouseButtonDown(0) && mainCamera !=null)
        {
            //从鼠标位置创建射线
            Ray ray=mainCamera.ScreenPointToRay(Input.mousePosition);
            if (Physics.Raycast(ray, out RaycastHit hitInfo))
            {
                //根据标签判断单击到的物体是否为地面
                if (hitInfo.collider.CompareTag("Ground"))
                    Debug.Log(string.Format("单击地面{0}", hitInfo.point));
            }
        }
    }
}
```

在鼠标单击到地面时，通常会在单击位置生成一个特效，用于标识单击位置，如图 4-1 所示。

该特效通过粒子系统实现，分为两部分，一部分是向下旋转的箭头，另一部分是从中间向外扩散的圆形，需要用到两张贴图，如图 4-2 所示。

箭头特效的渲染模式为 Mesh，用到一个弯曲的面片模型，如图 4-3 所示。

在 Color over Lifetime 中设置颜色渐变，使 Alpha 值在生命周期中从透明变为不透明，再从不透明变为透明，如图 4-4 所示。

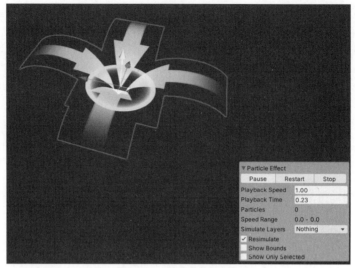

图 4-1　地面单击特效

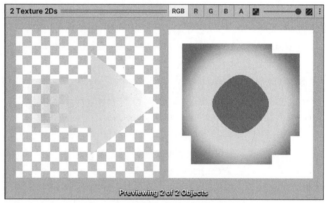

图 4-2　地面单击特效所用贴图

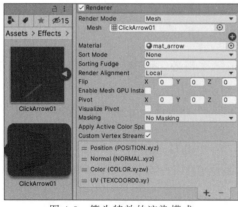

图 4-3　箭头特效的渲染模式

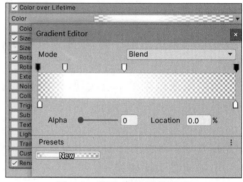

图 4-4　箭头特效颜色渐变

在 Rotation over Lifetime 中设置箭头的旋转方向，使箭头在生命周期中沿指定的轴进行旋转，如图 4-5 所示。

在 Size over Lifetime 中通过动画曲线设置箭头在生命周期中的大小，从一定规模扩大到最大值，再逐渐缩小，如图 4-6 所示。

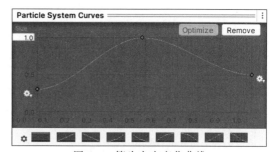

图 4-5　箭头旋转　　　　　　　　　图 4-6　箭头大小变化曲线

圆形特效与箭头特效的主要区别在于圆形特效的渲染模式为 Horizontal Billboard，在生命周期中的大小从 0 逐渐增长到 1。

当需要获取地面单击位置时，实例化特效，根据 hitInfo 中的碰撞位置和法线方向设置特效的坐标和朝向，最终播放粒子系统，代码如下：

```
//第 4 章/GroundClick.cs

Ray ray=mainCamera.ScreenPointToRay(Input.mousePosition);
if (Physics.Raycast(ray, out RaycastHit hitInfo))
{
    //根据标签判断单击到的物体是否为地面
    if (hitInfo.collider.CompareTag("Ground"))
    {
        Debug.Log(string.Format("单击地面{0}", hitInfo.point));
        //实例化特效
        var instance=Instantiate(effect);
        //将特效坐标设为碰撞位置加上方一定单位
        instance.transform.position=hitInfo.point +Vector3.up * .1f;
        //根据法线方向设置特效的上方
        instance.transform.up=hitInfo.normal;
        //播放
        instance.Play();
    }
}
```

特效播放完成后需要将其销毁，可以通过协程实现一个延时销毁组件，将其挂载于特效物体即可，代码如下：

```csharp
//第 4 章/DelayDestroy.cs

using UnityEngine;
using System.Collections;

public class DelayDestroy : MonoBehaviour
{
    [SerializeField]
    private float delay=3f;

    private IEnumerator Start()
    {
        //延时销毁
        yield return new WaitForSeconds(delay);
        Destroy(gameObject);
    }
}
```

4.1.2 游戏物体事件响应系统

假如需要通过鼠标与场景中的游戏物体进行交互,可以使用 MonoBehaviour 中相关的回调方法为游戏物体添加碰撞器后,当鼠标有进入、单击或退出碰撞器等行为时,相关的回调方法将被调用,详解见表 4-2。

表 4-2 MonoBehaviour 中与碰撞器交互相关的回调方法

方法	详解
OnMouseEnter()	当鼠标进入碰撞器时被调用
OnMouseOver()	当鼠标悬浮在碰撞器上时每帧被调用
OnMouseDown()	当鼠标在碰撞器上单击时被调用
OnMouseDrag()	当鼠标在碰撞器上按下并保持按住时被调用
OnMouseUpAsButton()	当鼠标在同一碰撞器上按下并松开时被调用
OnMouseExit()	当鼠标退出碰撞器时被调用

通过这些 MonoBehaviour 中的回调方法与游戏物体进行交互的方式适用于简单的场景,当场景中的碰撞器变得复杂时,此方式将不再合适。

例如,场景中的某个区域是不可进入的,当角色移动到该区域边缘时将被无形的碰撞器阻挡,而该区域内的某些游戏物体需要通过鼠标单击进行交互,此时鼠标单击到的将是该区域的无形的碰撞器,区域内游戏物体的碰撞器无法被单击,回调方法也就不会被调用。

在这种情况下,便需要使用射线投射检测自定义一个事件响应系统,系统中将通过设置 LayerMask 实现只检测指定层级中的碰撞器。

首先定义事件响应器接口,事件响应器将挂载于与鼠标有交互行为的碰撞器所对应的游戏物体,代码如下:

```csharp
//第 4 章/IEventResponser.cs

//<summary>
//事件响应器接口
//</summary>
public interface IEventResponser
{
    //<summary>
    //进入事件
    //</summary>
    void OnEnter();
    //<summary>
    //单击事件
    //</summary>
    void OnClick();
    //<summary>
    //停留事件
    //</summary>
    void OnStay();
    //<summary>
    //退出事件
    //</summary>
    void OnExit();
}
```

在事件响应系统中声明一个 IEventResponser 类型字段,用于记录当前检测到的事件响应器。在 Update() 方法中,从鼠标位置创建射线进行射线投射检测,当检测到碰撞器时,判断该碰撞器是否挂载事件响应器组件。如果挂载,首先执行当前事件响应器的退出事件,然后将当前事件响应器更新为最新检测到的事件响应器,更新后执行当前事件响应器的进入事件,如果物体未挂载事件响应器组件或者射线投射检测未检测到任何碰撞器,并且当前事件响应器不为空,则执行其退出事件并置为空,代码如下:

```csharp
//第 4 章/EventResponseSystem.cs

using UnityEngine;
```

```csharp
//<summary>
//事件响应系统
//</summary>
public class EventResponseSystem : MonoBehaviour
{
    //主相机
    [SerializeField]
    private Camera mainCamera;
    //启用时自动查找相机（当主相机为空时起作用）
    [SerializeField]
    private bool autoFindCameraOnEnable=true;
    //检测层级
    [SerializeField]
    private LayerMask eventResponserLayerMask=-1;
    //检测的最大距离
    [SerializeField]
    private float maxDistance=10f;

    //<summary>
    //当前事件响应器
    //</summary>
    public IEventResponser CurrentEventResponser { get; private set; }

    private void OnEnable()
    {
        if (mainCamera==null && autoFindCameraOnEnable)
            //如果 Tag 为 MainCamera 的主相机不存在，则根据类型查找相机
            mainCamera=Camera.main !=null ? Camera.main
                : FindObjectOfType<Camera>();
    }

    private void Update()
    {
        if (mainCamera==null) return;

        //在鼠标位置进行射线投射检测
        Ray ray=mainCamera.ScreenPointToRay(Input.mousePosition);
        if (Physics.Raycast(ray, out RaycastHit hit,
            maxDistance, eventResponserLayerMask))
        {
            //检测到的碰撞器挂载事件响应器组件
```

```csharp
                var eventResponser=hit.collider
                    .GetComponent<IEventResponser>();
                if (eventResponser !=null)
                {
                    //首先执行当前的事件响应器的退出事件
                    CurrentEventResponser?.OnExit();
                    //更新当前的事件响应器
                    CurrentEventResponser=eventResponser;
                    //更新后执行当前的事件响应器的进入事件
                    CurrentEventResponser.OnEnter();
                }
                //检测到的碰撞器未挂载事件响应器
                else
                {
                    //如果当前事件响应器不为空,则执行其退出事件并置为空
                    if (CurrentEventResponser !=null)
                    {
                        CurrentEventResponser.OnExit();
                        CurrentEventResponser=null;
                    }
                }
            }
            //未检测到任何碰撞器
            else
            {
                //如果当前事件响应器不为空,则执行其退出事件并置为空
                if (CurrentEventResponser !=null)
                {
                    CurrentEventResponser.OnExit();
                    CurrentEventResponser=null;
                }
            }
        }

        private void OnDisable()
        {
            if (CurrentEventResponser !=null)
            {
                CurrentEventResponser.OnExit();
                CurrentEventResponser=null;
            }
```

```
        }
    }
```

如果当前事件响应器不为空,则每帧执行其停留事件,鼠标单击时执行其单击事件,代码如下:

```
//第 4 章/EventResponseSystem.cs

//当前事件响应器不为空
if (CurrentEventResponser !=null)
{
    //执行其停留事件
    CurrentEventResponser.OnStay();

    //鼠标单击执行其单击事件
    if (Input.GetMouseButtonDown(0))
        CurrentEventResponser.OnClick();
}
```

接下来实现一个事件响应器的具体示例,Outline 组件用于实现物体边缘高亮效果,当鼠标进入事件响应器所在的游戏物体时,激活边缘高亮效果,并通过一个文本组件显示该游戏物体的描述,当鼠标退出事件响应器所在的游戏物体时,取消激活边缘高亮效果,并隐藏文本,代码如下:

```
//第 4 章/EventResponserSample.cs

using UnityEngine;
using UnityEngine.UI;
using Outline=QuickOutline.Outline;

public class EventResponserSample : MonoBehaviour, IEventResponser
{
    [SerializeField]
    protected Outline outline;
    [SerializeField]
    protected string description;
    [SerializeField]
    private Text descriptionText;

    public void OnEnter()
    {
```

```
            outline.enabled=true;
            descriptionText.text=description;
            descriptionText.gameObject.SetActive(true);
        }
        public void OnExit()
        {
            outline.enabled=false;
            descriptionText.gameObject.SetActive(false);
        }
        public void OnStay()
        {
            descriptionText.rectTransform
                .anchoredPosition3D=Input.mousePosition;
        }
        public void OnClick() { }
}
```

为场景中的两个游戏物体分别添加该事件响应器组件,运行程序,通过鼠标与这些游戏物体进行交互,效果如图 4-7 所示。

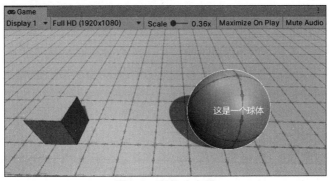

图 4-7　事件响应器

4.2　球体投射检测

球体投射检测使用 Physics 类中的 SphereCast()方法,球体投射检测方法相较于射线投射检测方法多了一个 radius 参数,该参数为 float 类型,用于指定球体的半径,代码如下:

```
public static bool SphereCast(Vector3 origin, float radius,
    Vector3 direction, out RaycastHit hitInfo);
public static bool SphereCast(Vector3 origin, float radius,
    Vector3 direction, out RaycastHit hitInfo, float maxDistance);
```

```csharp
public static bool SphereCast(Vector3 origin, float radius,
    Vector3 direction, out RaycastHit hitInfo,
    float maxDistance, int layerMask);
public static bool SphereCast(Vector3 origin, float radius,
    Vector3 direction, out RaycastHit hitInfo,
    float maxDistance, QueryTriggerInteraction queryTriggerInteraction);
public static bool SphereCast(Ray ray, float radius);
public static bool SphereCast(Ray ray, float radius, float maxDistance);
public static bool SphereCast(Ray ray, float radius, float maxDistance,
    int layerMask);
public static bool SphereCast(Ray ray, float radius, float maxDistance,
    int layerMask, QueryTriggerInteraction queryTriggerInteraction);
public static bool SphereCast(Ray ray, float radius,
    out RaycastHit hitInfo);
public static bool SphereCast(Ray ray, float radius,
    out RaycastHit hitInfo, float maxDistance);
public static bool SphereCast(Ray ray, float radius,
    out RaycastHit hitInfo, float maxDistance, int layerMask);
public static bool SphereCast(Ray ray, float radius,
    out RaycastHit hitInfo, float maxDistance, int layerMask,
    QueryTriggerInteraction queryTriggerInteraction);
```

2.2节介绍了第三人称视角相机的实现,第三人称视角相机用于跟随用户的人物角色,相机与角色之间有一定距离。如果相机与角色之间有其他游戏物体,则视角将会被阻挡,导致在视角中无法看到用户的人物角色,因此在第三人称视角相机组件中,通常会实现避障功能,目标是使相机避开与人物角色之间的障碍物,始终使人物角色保持在画面中。

相机避障功能可以通过球体投射检测实现,从角色的位置向相机所在位置进行球体投射检测,如果检测到碰撞器,则说明相机与人物角色之间存在障碍物,将碰撞点沿射线反方向减去一定单位即可获得避障后的坐标点,代码如下:

```csharp
//第 4 章 / ThirdPersonCameraController.cs

//碰撞信息
private RaycastHit hitInfo;

//避障
private Vector3 ObstacleAvoidance(Vector3 current,
    Vector3 target, float radius, float maxDistance, LayerMask layerMask)
{
    Ray ray=new Ray(target, current-target);
```

```csharp
    if (Physics.SphereCast(ray, radius,
        out hitInfo, maxDistance, layerMask))
    {
        return ray.GetPoint(hitInfo.distance-radius * 2f);
    }
    return current;
}
```

通过 Gizmos 类和 Handles 类中的方法在场景中绘制出球体投射检测的起点和终点，以及检测到障碍物时的碰撞点和避障后的点，实现可视化调试，代码如下：

```csharp
//第 4 章 / ThirdPersonCameraController.cs

//相机半径
[SerializeField] private float cameraRadius=.2f;
//障碍物层级
[SerializeField] private LayerMask obstacleLayer;

private void OnDrawGizmos()
{
    Vector3 current=transform.position;
    Vector3 target=avatar.position +Vector3.up * height;
    Vector3 final=ObstacleAvoidance(current, target,
        cameraRadius, distance, obstacleLayer);
    Gizmos.DrawLine(current, target);
    Gizmos.DrawWireSphere(current +Vector3.up * .1f, .1f);
    Gizmos.DrawWireSphere(target +Vector3.up * .1f, .1f);
    Gizmos.color=Color.yellow;
    Gizmos.DrawWireSphere(hitInfo.point +Vector3.up * .1f, .1f);
    Gizmos.color=Color.cyan;
    Gizmos.DrawWireSphere(final +Vector3.up * .1f, .1f);
#if UNITY_EDITOR
    UnityEditor.Handles.Label(target, "球形投射检测起点");
    UnityEditor.Handles.Label(current, "球形投射检测终点");
    UnityEditor.Handles.Label(hitInfo.point, "碰撞点");
    UnityEditor.Handles.Label(final, "避障后的点");
#endif
}
```

在场景中添加一个障碍物，将人物移动到该障碍物前方，此时视角被障碍物阻挡，结果如图 4-8 所示。

最终将相机的坐标设为避障后的点即可实现避障功能，代码如下：

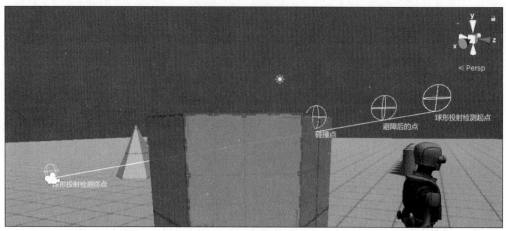

图 4-8　避障功能调试

```
//第 4 章/ ThirdPersonCameraController.cs

private void LateUpdate()
{
    //...
    Vector3 targetPosition=targetRotation * Vector3.back
        * targetDistance +avatar.position +Vector3.up * height;
    //避障
    targetPosition=ObstacleAvoidance(
        targetPosition,
        avatar.position +Vector3.up * height,
        cameraRadius,
        distance,
        obstacleLayer);
    //赋值
    transform.rotation=targetRotation;
    transform.position=targetPosition;
}
```

4.3　盒体重叠检测

　　Physics 类中不仅包含投射类型的物理检测方法，还包含重叠类型的物理检测方法，以盒体重叠检测方法为例，它用于检测与盒体接触或位于盒体内部的所有碰撞器，方法同样具有多个重载，代码如下：

```
public static Collider[] OverlapBox(Vector3 center, Vector3 halfExtents);
public static Collider[] OverlapBox(Vector3 center, Vector3 halfExtents,
    Quaternion orientation);
public static Collider[] OverlapBox(Vector3 center, Vector3 halfExtents,
    Quaternion orientation, int layerMask);
public static Collider[] OverlapBox(Vector3 center, Vector3 halfExtents,
    Quaternion orientation, int layerMask,
    QueryTriggerInteraction queryTriggerInteraction);
```

参数详解见表 4-3。

表 4-3　OverlapBox()方法参数详解

参　　数	详　　解
center	盒体的中心点
halfExtents	盒体各维度大小的一半
orientation	盒体的旋转
layerMask	用于指定盒体检测的层级
queryTriggerInteraction	用于指定盒体检测是否命中触发器

在 RTS 类型的游戏中通常会有通过鼠标框选游戏战斗单位的功能，该功能可以使用盒体重叠检测方法实现。首先需要在屏幕中绘制框选的范围，然后根据框选范围定位该范围在世界空间中对应的区域，最终在该区域内进行盒体重叠检测，检测区域内的游戏战斗单位。

框选范围可以通过 LineRenderer 组件进行绘制，该组件会根据指定的点位集合依次在两点之间绘制连线。Loop 属性表示首尾两点是否相连，形成闭环，将其设置为 true，并且将线条的宽度设为合适的大小，如图 4-9 所示。

因为框选范围是一个矩形，有 4 个顶点，所以将 LineRenderer 中的点位数量设为 4。需要注意的是，这里的点位坐标是在世界空间中的坐标，因此在确定框选范围后，需要将 4 个顶点从屏幕坐标转换为世界坐标。

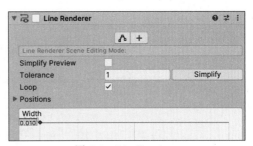

图 4-9　LineRenderer

在鼠标按下的那一帧中依据鼠标坐标位置生成矩形的第 1 个顶点，记为 P1，在鼠标拖曳过程中，生成矩形的第 2 个顶点，记为 P2，因为另外两个顶点与 P1 或 P2 具有相同的 x 值或 y 值，所以另外两个顶点可以通过 P1 和 P2 得知，代码如下：

```csharp
//第 4 章/BoxSelect.cs

using UnityEngine;

public class BoxSelect : MonoBehaviour
{
    [SerializeField]
    private Camera mainCamera;
    [SerializeField]
    private LineRenderer lineRenderer;
    private Vector3 screenP1;
    private Vector3 screenP2;

    private void Update()
    {
        //鼠标左键单击
        if (Input.GetMouseButtonDown(0))
        {
            //激活 LineRenderer 组件,开始绘制
            lineRenderer.enabled=true;
            //根据鼠标位置获得第 1 个顶点的坐标
            screenP1=Input.mousePosition;
            screenP1.z=1f;
        }
        //鼠标拖曳过程中
        if (Input.GetMouseButton(0))
        {
            //根据鼠标位置获得第 2 个顶点的坐标
            screenP2=Input.mousePosition;
            screenP2.z=1f;
            //另外两个顶点的坐标通过 P1 和 P2 获得
            Vector3 screenP3=new Vector3(screenP2.x, screenP1.y, 1f);
            Vector3 screenP4=new Vector3(screenP1.x, screenP2.y, 1f);
            //屏幕坐标转世界坐标 设置到 LineRenderer 中
            lineRenderer.SetPosition(0,
                mainCamera.ScreenToWorldPoint(screenP1));
            lineRenderer.SetPosition(1,
                mainCamera.ScreenToWorldPoint(screenP3));
            lineRenderer.SetPosition(2,
                mainCamera.ScreenToWorldPoint(screenP2));
            lineRenderer.SetPosition(3,
```

```csharp
            mainCamera.ScreenToWorldPoint(screenP4));
    }
    //鼠标左键抬起 框选结束
    if (Input.GetMouseButtonUp(0))
    {
        //取消激活 LineRenderer 组件
        lineRenderer.enabled=false;
    }
}
```

在世界空间中的顶点可以在鼠标按下和抬起时通过射线投射检测获得,代码如下：

```csharp
//第 4 章/BoxSelect.cs
private Vector3 worldP1;
private Vector3 worldP2;
private RaycastHit hitInfo;

private void Update()
{
    //鼠标左键单击
    if (Input.GetMouseButtonDown(0))
    {
        //激活 LineRenderer 组件,开始绘制
        lineRenderer.enabled=true;
        //根据鼠标位置获得第 1 个顶点的坐标
        screenP1=Input.mousePosition;
        screenP1.z=1f;
        //通过射线投射检测获取世界空间中的第 1 个顶点
        if (Physics.Raycast(mainCamera.ScreenPointToRay(
            Input.mousePosition), out hitInfo,
            100f, 1<<LayerMask.NameToLayer("Ground")))
        {
            worldP1=hitInfo.point;
        }
    }
    //...
    //鼠标左键抬起 框选结束
    if (Input.GetMouseButtonUp(0))
    {
```

```csharp
    //取消激活 LineRenderer 组件
    lineRenderer.enabled=false;
    //通过射线投射检测获取世界空间中的第 2 个顶点
    if (Physics.Raycast(mainCamera.ScreenPointToRay(
        Input.mousePosition), out hitInfo,
        100f, 1<<LayerMask.NameToLayer("Ground")))
    {
        worldP2=hitInfo.point;
    }
  }
}
```

计算出盒体的中心点和盒体各维度大小的一半,然后进行盒体重叠检测。向示例场景中的游戏战斗单位挂载边缘高亮效果组件,如果将该组件激活,则表示游戏战斗单位被选中。在框选时还需要清除上一次框选的内容,因此需要将框选的结果通过一个列表进行缓存,代码如下:

```csharp
//第 4 章/BoxSelect.cs

//缓存列表
private readonly List<Outline> cacheList=new List<Outline>();
private void Update()
{
    //...
    //鼠标左键抬起 框选结束
    if (Input.GetMouseButtonUp(0))
    {
        //取消激活 LineRenderer 组件
        lineRenderer.enabled=false;
        //清除上一次框选内容
        for (int i=0; i<cacheList.Count; i++)
            cacheList[i].enabled=false;
        cacheList.Clear();
        //通过射线投射检测获取世界空间中的第 2 个顶点
        if (Physics.Raycast(mainCamera.ScreenPointToRay(
            Input.mousePosition), out hitInfo,
            100f, 1<<LayerMask.NameToLayer("Ground")))
        {
            worldP2=hitInfo.point;
            //盒体的中心
            Vector3 center=new Vector3(
```

```
            (worldP1.x +worldP2.x) * .5f,
            .5f,
            (worldP1.z +worldP2.z) * .5f);
        //盒体各维度大小的一半
        Vector3 halfExtents=new Vector3(
            Mathf.Abs(worldP2.x-worldP1.x) * .5f,
            .5f,
            Mathf.Abs(worldP2.z-worldP1.z) * .5f);
        //盒体重叠检测
        Collider[] colliders=Physics.OverlapBox(center, halfExtents);
        for (int i=0; i<colliders.Length; i++)
        {
            if (colliders[i].TryGetComponent<Outline>(out var outline))
            {
                outline.enabled=true;
                cacheList.Add(outline);
            }
        }
    }
}
```

最终效果如图 4-10 所示。

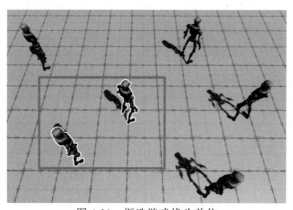

图 4-10　框选游戏战斗单位

4.4　物理检测可视化

在游戏开发过程中,物理检测是一个至关重要的环节,它涉及游戏角色与环境的交互、碰撞响应及物体行为的模拟等多个方面。为了更好地调试和优化物理检测过程,将其可视化成为一种常见的做法。本节将详细介绍如何实现物理检测的可视化，帮助开发者直观地查看和调试物理交互，提升游戏开发的效率和品质。

4.4.1 盒体重叠检测可视化

在 MonoBehaviour 的 OnDrawGizmos() 或 OnDrawGizmosSelected() 方法中，调用 Gizmos 类中的 DrawWireCube() 方法可以绘制立方体线框，代码如下：

```
public static void DrawWireCube(Vector3 center, Vector3 size);
```

假设在盒体重叠检测时未设置朝向参数，那么使用该方法可以绘制出盒体重叠检测的区域，实现可视化调试，但是当设置了朝向参数时，直接使用该方法便不再合适，因为它只支持设置立方体的中心点和大小，也就是说，该方法绘制出的立方体是一个轴对齐的立方体。

要绘制一个有向的立方体，可以通过在修改 Gizmos 类中表示矩阵的变量 matrix 后，再调用 DrawWireCube() 方法实现，也可以通过计算立方体的 8 个顶点，在顶点之间绘制连线实现。

本节使用第 2 种方式，实现一个绘制盒体线框的工具类，使用该类中的方法绘制出的盒体线框支持设置朝向。在绘制盒体时，首先根据中心点和大小确定盒体的 8 个顶点，中心点向各个轴向上偏移各轴向对应大小的一半便是顶点坐标，然后将顶点坐标根据朝向和变换矩阵进行变换，代码如下：

```
//第 4 章/PhysicsGizmos.cs
private static Vector3[] GetBoxVertices(Vector3 center,
Vector3 halfExtents, Quaternion orientation, Matrix4x4 world2Local)
{
    Vector3[] verts=new Vector3[8];
    int index=-1;
    for (int z=-1; z<=1; z+=2)
    {
        for (int y=-1; y<=1; y+=2)
        {
            for (int x=-1; x<=1; x+=2)
            {
                verts[++index]=center +new Vector3(
                    -Mathf.Sign(y) * x * halfExtents.x,
                    y * halfExtents.y,
                    z * halfExtents.z);
                verts[index]=orientation * world2Local
                    .MultiplyPoint(verts[index]);
                verts[index]=world2Local.inverse
```

```
            .MultiplyPoint(verts[index]);
        }
    }
}
    return verts;
}
```

确定盒体的 8 个顶点后,通过 Gizmos 类中的 DrawLine()方法在指定的两点之间绘制盒体的连线,代码如下:

```
//第 4 章/PhysicsGizmos.cs
public static void OverlapBoxWire(Vector3 center, Vector3 halfExtents,
    Quaternion orientation, Matrix4x4 world2Local)
{
    Vector3[] verts=GetBoxVertices(
        center, halfExtents, orientation, world2Local);
#if UNITY_EDITOR
    for (int i=0; i<verts.Length; i++)
        Handles.Label(verts[i], i.ToString(), m_style);
#endif
    for (int i=0; i<2; i++)
    {
        for (int j=0; j<4; j++)
        {
            Vector3 curr=verts[j %4 +i * 4];
            Vector3 next=verts[(j +1) %4 +i * 4];
            Gizmos.DrawLine(curr, next);
            if (i==0)
                Gizmos.DrawLine(curr, verts[j %4 +(i +1) * 4]);
        }
    }
}
```

注意 OverlapBoxWire()方法的第 2 个参数是指盒体大小的一半,假设盒体的大小是(0.8,0.8,0.8),那么该参数值为(0.4,0.4,0.4),示例代码如下:

```
private void OnDrawGizmos()
{
    PhysicsGizmos.OverlapBoxWire(transform.position +transform.up, .4f *
        Vector3.one, transform.rotation, transform.worldToLocalMatrix);
}
```

结果如图 4-11 所示。

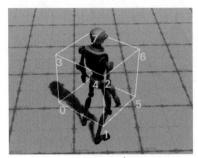

图 4-11 绘制盒体线框

4.4.2 盒体投射检测可视化

如果想要实现盒体投射检测的可视化，则需要绘制两个盒体，根据中心点和大小形成第 1 个盒体，根据投射的方向和最大距离生成第 2 个盒体。

需要划分为两个盒体是由于当碰撞进入第 1 个盒体区域时，投射检测方法的返回值将为 false，例如，以角色位置为中心点向其前方进行盒体投射检测，盒体的大小为(0.4，0.4，0.6)，最大距离设为 1.5，当检测到碰撞时，通过绘制球体标记碰撞位置，代码如下：

```
//第 4 章/PhysicsExample.cs
using UnityEngine;
#if UNITY_EDITOR
using UnityEditor;
#endif

public class PhysicsExample : MonoBehaviour
{
    [SerializeField]
    private Vector3 boxSize=new Vector3(.4f, .4f, .6f);
    [SerializeField]
    private float maxDistance=1.5f;
    private Vector3 center;
    private bool flag;
    private RaycastHit hitInfo;

    private void Update()
    {
        center=transform.position +transform.up;
        flag=Physics.BoxCast(center, boxSize * .5f, transform.forward,
            out hitInfo, transform.rotation, maxDistance);
```

```
    }
    private void OnDrawGizmos()
    {
        PhysicsGizmos.BoxCastWire(center, boxSize * .5f,
            transform.rotation, transform.worldToLocalMatrix,
            transform.forward, maxDistance);
        if (flag)
        {
            Gizmos.color=Color.red;
            Gizmos.DrawSphere(hitInfo.point, .05f);
#if UNITY_EDITOR
            Handles.Label(hitInfo.point,
                string.Format("碰撞点:{0} 碰撞距离:{1}",
                    hitInfo.point, hitInfo.distance),
                new GUIStyle(GUI.skin.label) { fontSize=15 });
#endif
        }
    }
}
```

当碰撞进入第 2 个盒体区域时,盒体投射检测结果为 true,如图 4-12(a)所示,当碰撞进入第 1 个盒体区域时,盒体投射检测结果为 false,如图 4-12(b)所示。

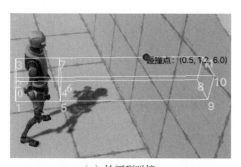 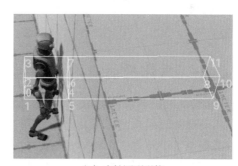

(a) 检测到碰撞　　　　　　　　　　(b) 未检测到碰撞

图 4-12　盒体投射检测结果

最大距离是从第 2 个盒体开始向投射方向延伸的距离,在 4.4.1 节中介绍了如何根据中心点和大小获取盒体的 8 个顶点,在此基础上,根据投射方向和最大距离增加 4 个顶点,并绘制相应连线,这样便可以实现盒体投射检测的可视化,代码如下:

```
//第 4 章/PhysicsGizmos.cs
public static void BoxCastWire(Vector3 center, Vector3 halfExtents,
```

```
        Quaternion orientation, Matrix4x4 world2Local,
        Vector3 direction, float maxDistance)
{
    Vector3[] verts=GetBoxVertices(
        center, halfExtents, orientation, world2Local);
    Vector3[] verts2=new Vector3[4];
    for (int i=0; i<verts2.Length; i++)
        verts2[i]=verts[i +4] +maxDistance * direction;
    verts=verts.Concat(verts2).ToArray();
#if UNITY_EDITOR
    for (int i=0; i<verts.Length; i++)
        Handles.Label(verts[i], i.ToString(), m_style);
#endif
    for (int i=0; i<3; i++)
    {
        for (int j=0; j<4; j++)
        {
            Vector3 curr=verts[j %4 +i * 4];
            Vector3 next=verts[(j +1) %4 +i * 4];
            Gizmos.DrawLine(curr, next);
            if (i==0)
                Gizmos.DrawLine(curr, verts[j %4 +(i +2) * 4]);
        }
    }
}
```

4.4.3 球体投射检测可视化

球体线框的绘制其实是在3个维度上绘制圆形,实现球体投射检测可视化需要绘制3个球体线框,在第2个与第3个球体线框的相应顶点之间绘制连线,形成一个胶囊体,如图4-13所示。当碰撞进入第1个球体区域时,投射检测方法的返回值为false,当碰撞进入胶囊体区域时,投射检测方法的返回值为true。

球体线框包含6个顶点,由中心点向各个轴向上偏移一个半径大小获得,代码如下:

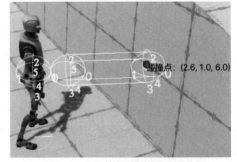

图4-13 盒体投射检测可视化

//第4章/PhysicsGizmos.cs

```csharp
private static Vector3[] GetSphereVertices(Vector3 center,
    float radius, Quaternion orientation, Matrix4x4 world2Local)
{
    List<Vector3>verts=new List<Vector3>();
    for (int z=1; z>=-1; z-=2)
        verts.Add(center +z * radius * Vector3.forward);
    for (int y=1; y>=-1; y-=2)
        verts.Add(center +y * radius * Vector3.up);
    for (int x=1; x>=-1; x-=2)
        verts.Add(center +x * radius * Vector3.right);
    for (int i=0; i<verts.Count; i++)
    {
        verts[i]=orientation * world2Local.MultiplyPoint(verts[i]);
        verts[i]=world2Local.inverse.MultiplyPoint(verts[i]);
    }
    return verts.ToArray();
}
```

球体线框的 3 个圆形可以通过 Handles 类中的 DrawWireDisc()方法绘制，通过设定中心点、法线方向和半径参数值即可绘制出圆形。因为两个向量的叉乘结果是垂直于这两个向量形成的平面的另一个向量，所以圆的法线方向可以通过向量叉乘计算获得，代码如下：

```csharp
//第 4 章/PhysicsGizmos.cs
private static void SphereWire(Vector3 center, Vector3[] verts, float radius)
{
#if UNITY_EDITOR
    for (int i=0; i<verts.Length; i++)
        Handles.Label(verts[i], i.ToString(), m_style);
#endif
    for (int i=0; i<6; i+=2)
    {
        Vector3 dir1=verts[(i +2) %6]-center;
        Vector3 dir2=verts[(i +4) %6]-center;
        Vector3 normal=Vector3.Cross(dir1, dir2);
#if UNITY_EDITOR
        Color cache=Handles.color;
        Handles.color=Gizmos.color;
        Handles.DrawWireDisc(center, normal, radius);
        Handles.color=cache;
#endif
```

```
    }
}
public static Vector3[] SphereWire(Vector3 center,
    float radius, Quaternion orientation, Matrix4x4 world2Local)
{
    Vector3[] verts=GetSphereVertices(
        center, radius, orientation, world2Local);
    SphereWire(center, verts, radius);
    return verts;
}
```

通过SphereWire()方法绘制出第1个球体线框后,由第1个球体线框的6个顶点,向球体投射检测的方向分别延伸球体直径和最大检测距离大小,这样便可以得到第2个和第3个球体线框的顶点坐标,在第2个与第3个球体线框的相应顶点坐标之间绘制连线,便可以形成胶囊体,代码如下:

```
//第4章/PhysicsGizmos.cs

public static void SphereCastWire(Vector3 origin,
float radius, Quaternion rotation, Matrix4x4 world2Local,
Vector3 direction, float maxDistance)
{
    Vector3[] verts1=SphereWire(
        origin, radius, rotation, world2Local);
    Vector3[] verts2=new Vector3[6];
    Vector3[] verts3=new Vector3[6];
    for (int i=0; i<6; i++)
    {
        verts2[i]=verts1[i] +radius * 2f * direction;
        verts3[i]=verts1[i] +maxDistance * direction;
        if (i>=2)
            Gizmos.DrawLine(verts2[i], verts3[i]);
    }
    SphereWire(origin +radius * 2f * direction, verts2, radius);
    SphereWire(origin +maxDistance * direction, verts3, radius);
}
```

第 5 章 动画系统

Unity 的动画系统基于动画剪辑(Animation Clip)的概念,动画剪辑包含特定对象应如何随时间改变其位置、旋转或其他相关属性的相关信息。Animator Controller 作为动画状态机,负责集中管理由动画剪辑对应的所有状态,负责跟踪当前应该播放哪个剪辑及动画应该何时改变或混合在一起。通过本章的学习,读者将能够熟练掌握 Animator 的使用方法和技巧,为创建生动形象的游戏动画打下坚实的基础。

5.1 动画剪辑

Unity 的动画系统基于关键帧动画的概念,多个关键帧构成了动画的连续变化,这些关键帧动画信息存储于动画剪辑资产中。

通过编辑动画剪辑,开发者可以创建、修改和管理其中的关键帧,从而实现角色、物体或其他游戏元素的动画效果。

动画剪辑在 Animation 窗口中进行编辑,该窗口通过菜单 Window/Animation/Animation 或者快捷键 Ctrl+6 打开,如图 5-1 所示。

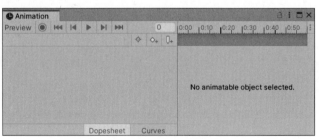

图 5-1 Animation 编辑窗口

开发者可以在时间轴上设置关键帧,然后系统会根据这些关键帧来自动计算中间帧,从而创建出平滑的动画效果。

关键帧可以通过手动创建,也可以通过进入录制模式录制关键帧,本节将介绍这两种关键帧创作方式。

5.1.1 录制关键帧

观察 Animation 窗口,可以看到左上角 Preview 按钮的右侧有一个红色的按钮,单击该按钮即可进入录制模式,如果在录制模式下对场景中的对象进行改动,系统则会自动在时间轴中的当前位置生成关键帧,记录修改的属性。

例如,在示例场景中创建一个 Cube 物体,为其创建一个新的 Animation Clip 资产,进入录制模式后,在 0:00 时间点将其坐标 x 值修改为 5,在 1:00 时间点将其坐标 x 值修改为 10,可以看到,系统自动在时间轴中的对应位置生成了关键帧,如图 5-2 所示。

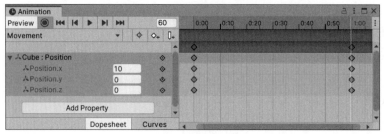

图 5-2 录制关键帧

录制完成后,再次单击录制按钮退出录制模式,单击"播放"按钮,即可进入预览状态,从而可以预览动画。

5.1.2 创建和编辑关键帧

在非录制模式下,如果要在指定的时间点记录对象属性,则需要先手动添加关键帧再进行记录,关键帧可以通过单击 Add keyframe 按钮进行添加,也可以在时间轴中右击并选择 Add Key 选项进行添加,如图 5-3 所示。

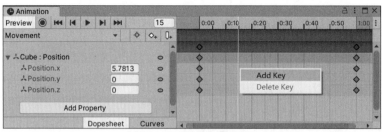

图 5-3 添加关键帧

添加关键帧之后可以在左侧的属性列表中编辑对应属性的值,也可以单击窗口下方的 Curves 按钮,进入曲线编辑窗口,通过编辑曲线的方式编辑属性值,如图 5-4 所示。

5.1.3 外部导入的动画资产

动画剪辑可以是在 Unity 中创建的资产,也可以是由外部导入的 fbx 文件包含的资产。

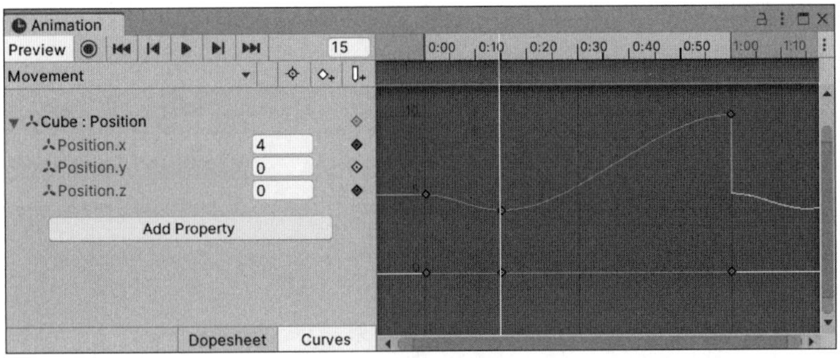

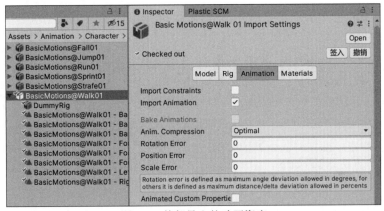

图 5-4　曲线编辑窗口

外部导入的动画剪辑可能包括美术工作人员在 3ds Max、Maya 或 Blender 中创建的动画，也可能是来自第三方库的动画集，如图 5-5 所示。

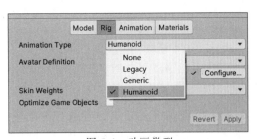

图 5-5　外部导入的动画资产

Import Animation 选项默认为勾选状态，表示导入外部动画，如果取消勾选，则动画资产将不会被导入。

打开 Rig 选项卡，可以看到动画类型有 None、Legacy、Generic 和 Humanoid 共 4 种类型，如图 5-6 所示。Legacy 是旧版本中使用的动画类型，已经不推荐使用。Generic 表示通用动画，Humanoid 表示人形动画，它们的核心区别在于，Humanoid 类型的动画可以通过骨骼重定向复用不同的人物角色动画，并且支持 IK 功能，IK 功能将在后续的章节中进行介绍。

图 5-6　动画类型

在 Animation 选项卡中可以对动画剪辑进行裁剪，如图 5-7 所示，Start 表示起始帧，End 表示结束帧。假设将 Start 设置为 0，将 End 设置为 15，表示截取 0～15 帧的动画。

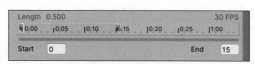

图 5-7 动画裁剪

5.2 动画状态机

动画状态机（Animator Controller）资产通过菜单 Assets/Create/Animator Controller 创建，双击该类型资产或者通过菜单 Window/Animation/Animator 可以打开 Animator 窗口，在该窗口中可以对动画状态机进行编辑。

5.2.1 Animator 窗口

Animator 窗口分为两个主要部分，分别是左侧的 Layers 与 Parameters 编辑区域和右侧的状态编辑区域。

动画参数在 Parameters 视图中进行编辑，包含 Float、Int、Bool 和 Trigger 共 4 种参数类型，如图 5-8 所示。在脚本中访问这些参数并向其赋值，即可实现状态间的切换，4 种类型参数对应的脚本中设置参数值的方法分别是 Animator 类中的 SetFloat()、SetInteger()、SetBool() 及 SetTrigger() 方法，代码如下：

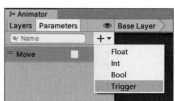

图 5-8 参数类型

```
public void SetFloat (string name, float value);
public void SetFloat (string name, float value, float dampTime, float deltaTime);
public void SetFloat (int id, float value);
public void SetFloat (int id, float value, float dampTime, float deltaTime);
public void SetInteger (string name, int value);
public void SetInteger (int id, int value);
public void SetBool (string name, bool value);
public void SetBool (int id, bool value);
public void SetTrigger (string name);
public void SetTrigger (int id);
```

可以看到，除了可以通过参数名称设置参数值外，还可以通过参数 id 设置参数值，参数 id 通过 Animator 类中的 StringToHash() 方法获取，代码如下：

```
public static int StringToHash(string name);
```

在 Layers 视图中可以创建不同的动画层，如图 5-9 所示，使用动画层可以管理身体的不同部位。例如当前有行走和射击两个动画，通过将行走动画放到下身层，将射击动画放到上身层，可以实现边行走边射击的动画融合效果。

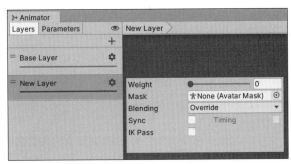

图 5-9　动画层

单击动画层右侧的齿轮按钮打开该层的设置弹窗，Weight 用于设置该动画层起作用的权重值，Avatar Mask 用于设定身体起作用的部位，该类型资产通过菜单 Assets/Create/Avatar Mask 进行创建。Blending 用于设置如何应用该动画层，包含 Override 和 Additive 两种类型，这决定了该层动画是替换上层动画还是叠加到上层动画。Sync 用于设置是否同步其他层的状态机，开启后，该层的状态机与要同步的层的状态机具有相同的结构，但是实际使用的动画剪辑可以不同。IK Pass 用于设置是否启用 IK。

5.2.2　动画状态

动画状态是动画状态机的基本组成部分，通过右击空白区域，选择 Create State/Empty 进行创建，如图 5-10 所示，也可以通过将动画剪辑拖入状态编辑区域进行创建。

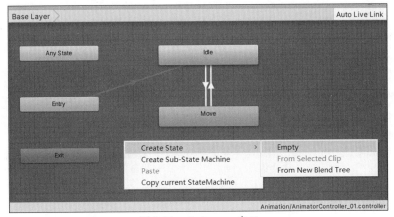

图 5-10　Animator 窗口

选中某个动画状态，在 Inspector 窗口中查看状态的属性，如图 5-11 所示，属性详解见表 5-1。

表 5-1 动画状态属性

属　　性	详　　解
Motion	该动画状态所引用的动画剪辑
Speed	该动画状态的默认播放速度,可以通过参数的形式进行修改
Motion Time	用于播放该动画状态的动作的时间
Mirror	是否启用镜像动画(仅用于人形动画),例如为一个向左行走的动画开启镜像,得到的是向右行走的动画
Cycle Offset	时间偏移量,当动画被设置为循环播放时,该属性将决定从动画的哪部分开始播放,例如当时间偏移量为 0 时表示从头开始播放,当时间偏移量为 0.5 时表示动画从其一半的位置开始播放
Foot IK	是否启用脚部 IK(仅用于人形动画)
Write Defaults	用于控制 Animator 是否将未动画化的属性写回其默认值
Transitions	源自该动画状态的过渡列表

当角色的行为十分复杂时,状态机将非常庞大,在这种情况下通常会使用子状态机对状态进行管理。子状态机通过在状态编辑区域的空白区域右击并选择 Create Sub-State Machine 进行创建,如图 5-12 所示。双击子状态机即可进入子状态机的编辑窗口,窗口顶部会显示当前正在编辑的子状态机。

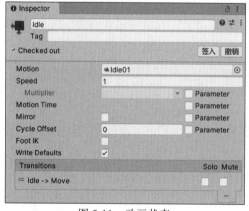

图 5-11　动画状态

图 5-12　Create Sub-State Machine

5.2.3　动画过渡

状态之间的带有箭头的连线是过渡线,表示状态过渡,单击过渡线,在 Inspector 窗口中编辑状态过渡的属性,如图 5-13 所示,这些属性决定了状态机从一种状态过渡到另一种状态的混合时长,以及应该在什么条件下触发过渡,例如从 Idle 状态过渡到 Move 状态的过渡条件是当 Move 参数变为 true 时。

动画过渡的属性详解见表 5-2。

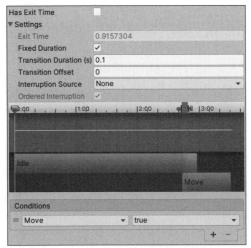

图 5-13 动画过渡

表 5-2 动画过渡属性

属　性	详　解
Has Exit Time	当动画过渡不依赖过渡条件时启用此选项,表示经过指定的标准化时间后进行过渡
Exit Time	当勾选 Has Exit Time 时,此值表示进行过渡的标准化时间
Fixed Duration	当值为 true 时表示过渡时间以秒为单位,当值为 false 时表示过渡时间以源状态的标准化时间进行表示
Transition Duration	过渡持续时间,当 Fixed Duration 为 true 时此值表示秒数,当 Fixed Duration 为 false 时此值表示标准化时间
Transition Offset	过渡到的目标状态的起始播放时间偏移值
Interruption Souce	控制过渡中断的类型
Ordered Interruption	确定当前过渡是否可在不考虑顺序的情况下被其他过渡中断,默认值为 true

Interruption Souce 的类型及详解见表 5-3。

表 5-3 Interruption Souce 详解

类　型	详　解
None	不添加任何过渡
Current State	将当前状态的过渡排队
Next State	对下一状态的过渡进行排队
Current State then Next State	将当前状态的过渡排队,然后将下一状态的过渡排队
Next State then Current State	将下一状态的过渡排队,然后将当前状态的过渡排队

如果将 Interruption Souce 设为 None,则在状态过渡期间不会因任何其他状态过渡的触发而中断,但是如果将其设为 Current State,则在状态过渡期间可能会因为源状态上的一些状态过渡触发而中断,为了更深入地理解,在示例状态机中添加 Jump、Talk 两个新状态,如图 5-14 所示,选中 Idle 状态,过渡优先级设置如图 5-15 所示。

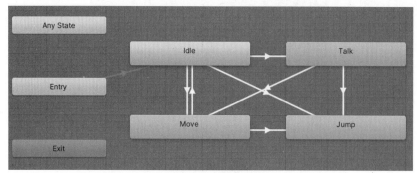

图 5-14 状态机

图 5-15 过渡优先级

在 Idle 至 Move 的状态过渡中,将 Interruption Souce 设为 Current State,那么在该状态过渡期间,如果 Idle 至 Talk 的状态过渡被触发,则 Idle 至 Move 的状态过渡将保持不间断,但是,如果 Idle 至 Jump 的状态过渡被触发,则 Idle 至 Move 的状态过渡将立即中断,开始向 Jump 状态过渡。

当然,以上是在 Ordered Interruption 属性为 true 的前提下,如果 Ordered Interruption 属性为 false,则 Idle 至 Talk、Idle 至 Jump 均会中断 Idle 至 Move 的状态过渡,如果它们在同一帧中触发,则状态会向 Jump 过渡,因为 Idle 至 Jump 具有更高的优先级。

如果将 Interruption Souce 设置为 Next State,则 Idle 至 Talk、Idle 至 Jump 不会中断 Idle 至 Move 的状态过渡,但是当 Move 至 Jump 的状态过渡被触发时,Idle 将立即开始向 Jump 状态过渡。

Current State then Next State、Next State then Current State 的类型则是为了更好地控制,在这种情况下,状态机将先分析前一种状态的过渡再分析后一种状态的过渡。例如当设置为 Current State then Next State 时,在 Idle 至 Move 的状态过渡期间,如果 Idle 至 Talk、Idle 至 Jump、Move 至 Jump 的状态过渡同时被触发,则状态机会开始向 Jump 状态过渡。

5.2.4 混合树

混合树用于实现多个动作之间的平滑混合,每个动作对最终效果的影响由混合参数决定。混合树通过在空白区域右击并选择 Create State/From New Blend Tree 进行创建。

例如,当前有 Idle、Walk、Run 共 3 个动画,分别是角色的静止动画、行走动画和奔跑动画,为了取得从静止到走跑或者从走跑到静止的良好的混合效果,将它们放在一个混合树中,通过一个 float 类型的参数进行控制,如图 5-16 所示。

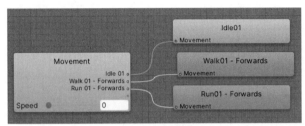

图 5-16　混合树

选中混合树,可以在检视窗口为各个动画设置阈值,如图 5-17 所示,当参数为 0 时,角色将处于静止状态,当参数从 0 逐渐过渡到 2 时,角色也将从静止动画逐渐过渡到行走动画,当参数逐渐过渡到 3.75 时,角色动画也将逐渐过渡到奔跑动画。

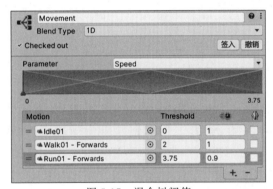

图 5-17　混合树阈值

以上是 1D 混合类型的混合树,假设角色有不同方向的移动动画,例如向前走、向左前方走、向右前方走等,移动方向需要通过两个参数进行控制,如图 5-18 所示,此时便需要 2D 混合类型的混合树。

2D 混合类型有多种,包括 2D Simple Directional、2D Freefoom Directional 和 2D Freeform Cartesian,它们具有不同的用途。

2D Simple Directional 适用于运动表示不同的方向,但是在同一方向上不会有多个运动的情况。假设在一个方向上有多个运动,例如向前移动有向前走和向前跑,这种情况适合使用 2D Freeform Directional 混合类型。2D Freeform Cartesian 适用于运动不表示不同方向的情况,两个参数表示不同的概念,例如速度和角速度。

Pos X 和 Pos Y 的取值可以通过 Compute Positions 下拉菜单选择不同的方式进行计算,如图 5-19 所示。Velocity XZ 表示根据运动的 Velocity X 和 Velocity Z 分别设置 Pos X

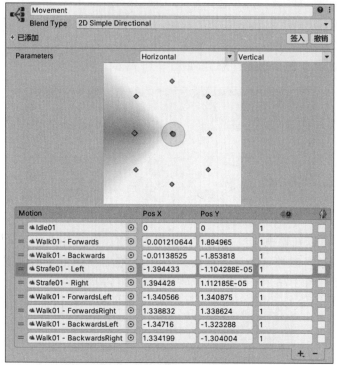

图 5-18 2D Simple Directional

和 Pos Y，Speed And Angular Speed 表示根据运动的角速度（弧度/秒）和速度分别设置 Pos X 和 Pos Y，而 X Position From 表示仅根据其中一项设置 Pos X，PosY 保持不变，Y Position From 同理。

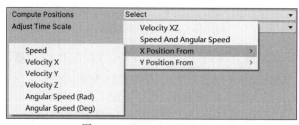

图 5-19 Compute Positions

5.3 动画事件

动画事件通常用于在动画播放的特定时间点触发某些操作，例如播放声音、切换动画状态、调用函数等。动画事件的实现方式有多种，下面介绍两种动画事件的实现途径。

5.3.1 Animation Clip Event

选中包含动画剪辑的 fbx 文件,在检视窗口中选择 Animation 视图,在下方的 Events 折叠栏中可以为相应的动画剪辑添加动画事件。

例如,在人物角色行走过程中,当脚踩在地上时,为其根据地面材质生成相应的脚印或者播放脚步音效,如果要实现类似功能,则可以通过在 Events 中添加动画帧事件实现,如图 5-20 所示。

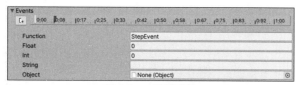

图 5-20 动画事件

需要注意的是,Function 用于设置事件需要调用的方法名,该方法需要在挂载 Animator 组件的游戏物体上的组件中实现,如果没有实现或者方法名不匹配,则在动画播放到相应位置时将会报错,如图 5-21 所示。

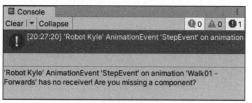

图 5-21 动画事件调用失败

Float、Int、String 及 Object 用于传递方法所需的参数值,如果方法没有参数,则不需要关心。

5.3.2 State Machine Behaviour

State Machine Behaviour 是指在状态机中为动画状态添加的动画行为脚本,通过选中状态,在检视面板中单击 Add Behaviour 按钮进行添加,如图 5-22 所示。重写 State Machine Behaviour 类中的虚方法即可实现相关的动画事件,详解见表 5-4。例如,进入 Talk 状态时,开始播放相应的音频剪辑,可以在 OnStateEnter() 方法中实现。

表 5-4 动画行为事件详解

方法	详解
OnStateEnter()	当状态进入时被调用
OnStateExit()	当状态退出时被调用
OnStateIK()	在 MonoBehaviour 的 OnAnimatorIK() 后被调用

续表

方法	详解
OnStateMove()	在 MonoBehaviour 的 OnAnimatorMove()后被调用
OnStateUpdate()	在状态持续期间的每个帧上被调用（除第 1 帧和最后一帧以外）

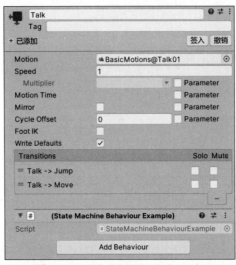

图 5-22　State Machine Behaviour

5.4　动画曲线

选中包含动画剪辑的 fbx 文件，在检视窗口中选择 Animation 视图，在下方的 Curves 折叠栏中可以为相应的动画剪辑添加动画曲线，如图 5-23 所示。

动画曲线的 x 轴表示动画剪辑的标准化时间，取值范围为 $[0,1]$，双击动画曲线便可打开曲线编辑器窗口，如图 5-24 所示，可以在曲线编辑器窗口中对曲线进行编辑。

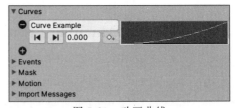

图 5-23　动画曲线

在 Animator 窗口中添加一个与动画曲线同名的 float 类型的参数，该参数值取自该曲线在时间轴中对应的值，在脚本中调用 Animator 类中的 GetFloat()方法，方法返回值等于方法调用时的曲线值，代码如下：

```
float value =animator.GetFloat("Curve Example");
```

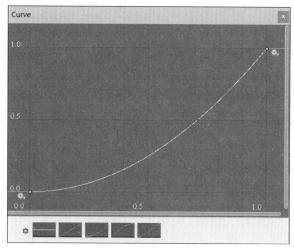

图 5-24　曲线编辑窗口

5.5　BlendShape

BlendShape 是一种基于关键帧动画的技术，主要用于实现面部表情动画，通过对模型的顶点进行形变实现表情变化。在 Unity 中，使用 Skinned Mesh Renderer 组件实现 BlendShape，如图 5-25 所示。

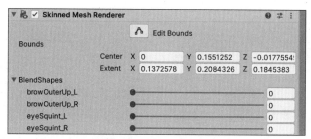

图 5-25　BlendShape

每个 BlendShape 对应一个表情变化，通过调节 BlendShape 的权重值，实现面部表情的变化。设置权重值的方法为 SetBlendShapeWeight()，调用该方法需要传入两个参数，第 1 个参数是 BlendShape 的索引值，第 2 个参数是要设置的权重值，索引值可以通过 Mesh 类中的 GetBlendShapeIndex() 方法获取，示例代码如下：

```
//第5章/BlendShapeExample.cs

using UnityEngine;

public class BlendShapeExample : MonoBehaviour
```

```
{
    private SkinnedMeshRenderer smr;
    private void Start()
    {
        smr=GetComponent<SkinnedMeshRenderer>();
    }
    private void OnGUI()
    {
        GUILayout.BeginHorizontal();
        GUILayout.Label("Jaw Open");
        //根据 Blend Shape 名称获取索引值
        int index=smr.sharedMesh.GetBlendShapeIndex("jawOpen");
        float weight=GUILayout.HorizontalSlider(
            smr.GetBlendShapeWeight(index), //获取当前权重值
            0f, 100f, GUILayout.Width(200f));
        //根据滑动条设置 Blend Shape 权重值
        smr.SetBlendShapeWeight(index, weight);
        GUILayout.EndHorizontal();
    }
}
```

在示例模型中，名为 jawOpen 的 BlendShape 用于控制嘴部的张开形变，当权重值为 0 时，如图 5-26(a)所示；当权重值为 50 时，如图 5-26(b)所示。

 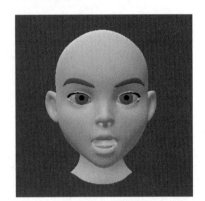

（a）jawOpen权重值为0　　　　　　（b）jawOpen权重值为50

图 5-26　调节 BlendShape 权重值

5.6　反向动力学

在正向动力学中，每个子关节的位置、角度都由父关节所支配，而反向动力学(Inverse Kinematics)是一种依据子关节的位置、角度反求父关节位置，从而确定整个骨架形态的方

法,以下简称 IK。

5.6.1 Animator IK

Animator 类中提供了设置 IK 坐标、旋转及设置 IK 权重值等相关方法,详解见表 5-5。

表 5-5 Animator 类中 IK 相关方法

IK 相关方法	详　　解
SetIKPosition()	设置反向动力学目标位置
SetIKPositionWeight()	设置反向动力学目标位置的权重值
SetIKRotation()	设置反向动力学目标旋转
SetIKRotationWeight()	设置反向动力学目标旋转的权重值
SetIKHintPosition()	设置反向动力学提示位置
SetIKHintPositionWeight()	设置反向动力学提示位置的权重值
GetIKPosition()	获取反向动力学目标位置
GetIKPositionWeight()	获取反向动力学目标位置的权重值
GetIKRotation()	获取反向动力学目标旋转
GetIKRotationWeight()	获取反向动力学目标旋转的权重值
GetIKHintPosition()	获取反向动力学提示位置
GetIKHintPositionWeight()	获取反向动力学提示位置的权重值

设置 IK 坐标、旋转的方法需要配合设置权重值的方法一起使用,当权重值为 0 时,设置的 IK 目标不起作用。设置 IK 的方法需要在 MonoBehaviour 中的 OnAnimatorIK() 回调方法中调用,该方法在即将更新其内部反向动力学系统前由动画组件调用。

需要注意的是,使用 IK 功能需要动画类型为 Humanoid 类型,即人形动画,如图 5-27 所示。当使用人形动画时,需要指定 Avatar,Avatar Definition 选项默认为 Create From

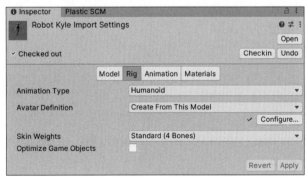

图 5-27 Humanoid Animation Type

This Model，表示从当前文件中定义的一组骨骼映射到人形 Avatar，也可以更改为 Copy From Other Avatar，表示使用其他模型文件定义的 Avatar，单击 Configure 按钮可以进入骨骼映射窗口，如图 5-28 所示，在该窗口中检查是否将模型的骨骼正确地映射到 Avatar。

另外，使用 IK 功能需要在 Animator 窗口中勾选对应动画层的 IK Pass 选项以启用 IK，否则 IK 不生效。为对应的动画层设置 Avatar Mask，以指定在该动画层起作用的身体部位，例如，只让上身部分起作用，Avatar Mask 的设置如图 5-29 所示。

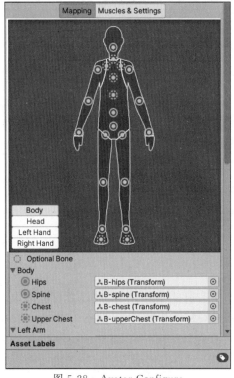

图 5-28 Avatar Configure

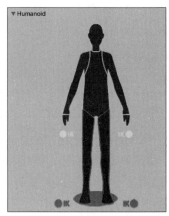

图 5-29 Avatar Mask

5.6.2　人物角色脚部放置方案

当人物角色处于静止状态时，由于受动画支配，在非平坦地面上角色的脚部并不贴合地面，如图 5-30 所示。为了使角色的脚部能够与不同类型的地面贴合，可以使用 IK 功能，为角色脚部设置 IK 的位置和旋转。

如何获取 IK 的位置和旋转？可以通过在脚部位置加上一定单位的高度处向下进行射线投射检测，碰撞信息中的碰撞点就是 IK 的目标位置，根据碰撞信息中的法线方向可以计算 IK 的目标旋转，代码如下：

图 5-30　人物角色脚部与地面不贴合情况

```
//第 5 章/FootIKExample.cs

//是否启用
[SerializeField] private bool enableFootIk=true;
//射线检测的长度
[SerializeField] private float raycastDistance=0.81f;
//射线检测的高度
[SerializeField] private float raycastOriginHeight=0.5f;
//左脚坐标
private Vector3 leftFootPosition;
//右脚坐标
private Vector3 rightFootPosition;
//左脚 IK 坐标
private Vector3 leftFootIkPosition;
//右脚 IK 坐标
private Vector3 rightFootIkPosition;
//左脚 IK 旋转
private Quaternion leftFootIkRotation;
//右脚 IK 旋转
private Quaternion rightFootIkRotation;
//左脚射线检测结果
private bool leftFootRaycast;
//右脚射线检测结果
private bool rightFootRaycast;

private void FixedUpdate()
{
```

```csharp
//未启用FootIK或者动画组件为空
if (!enableFootIk || animator==null) return;
#region 计算左脚IK目标
//左脚坐标
leftFootPosition=animator.GetBoneTransform(
    HumanBodyBones.LeftFoot).position;
leftFootPosition.y=transform.position.y+raycastOriginHeight;
//左脚射线投射检测
leftFootRaycast=Physics.Raycast(leftFootPosition,
    Vector3.down, out RaycastHit hit, raycastDistance, groundLayer);
if (leftFootRaycast)
{
    leftFootIkPosition=hit.point;
    leftFootIkRotation=Quaternion.FromToRotation(
        transform.up, hit.normal);
}
else leftFootIkPosition=Vector3.zero;
#endregion

#region 计算右脚IK目标
//右脚坐标
rightFootPosition=animator.GetBoneTransform(
    HumanBodyBones.RightFoot).position;
rightFootPosition.y=transform.position.y+raycastOriginHeight;
//右脚射线投射检测
rightFootRaycast=Physics.Raycast(rightFootPosition,
    Vector3.down, out hit, raycastDistance, groundLayer);
if (rightFootRaycast)
{
    rightFootIkPosition=hit.point;
    rightFootIkRotation=Quaternion.FromToRotation(
        transform.up, hit.normal);
}
else rightFootIkPosition=Vector3.zero;
#endregion
}
```

计算出IK目标位置和旋转后,在OnAnimatorIK()方法中进行应用,由于设置脚部IK位置会改变角色的高度,因此在设置IK前,需要先调整角色的身体高度,通过Animator类中的bodyPosition属性实现,代码如下:

```csharp
//第 5 章/FootIKExample.cs

//身体坐标插值速度
[Range(0f, 1f), SerializeField]
private float bodyPositionLerpSpeed=1f;
//脚部坐标插值速度
[Range(0f, 1f), SerializeField]
private float footPositionLerpSpeed=0.5f;
//左脚 Y 坐标缓存
private float lastLeftFootPositionY;
//右脚 Y 坐标缓存
private float lastRightFootPositionY;

private void OnAnimatorIK(int layerIndex)
{
    //未启用 FootIK 或者动画组件为空
    if (!enableFootIk || animator==null) return;

    //身体高度
    if (leftFootRaycast && rightFootRaycast)
    {
        //左脚坐标 Y 差值
        float leftPosYDelta=leftFootIkPosition.y -transform.position.y;
        //右脚坐标 Y 差值
        float rightPosYDelta=rightFootIkPosition.y -transform.position.y;
        //身体坐标 Y 差值取二者最小值
        float bodyPosYDelta=Mathf.Min(leftPosYDelta, rightPosYDelta);
        //目标身体坐标
        Vector3 targetBodyPosition=animator.bodyPosition
            +Vector3.up * bodyPosYDelta;
        //插值运算
        targetBodyPosition.y=Mathf.Lerp(animator.bodyPosition.y,
            targetBodyPosition.y, bodyPositionLerpSpeed);
        //设置身体坐标
        animator.bodyPosition=targetBodyPosition;
    }
    //应用左脚 IK 目标
    animator.SetIKPositionWeight(AvatarIKGoal.LeftFoot, 1f);
    animator.SetIKRotationWeight(AvatarIKGoal.LeftFoot, 1f);
    Vector3 targetIkPosition=animator.GetIKPosition(
        AvatarIKGoal.LeftFoot);
```

```csharp
        if (leftFootRaycast)
        {
            Vector3 world2Local=transform.InverseTransformPoint(
                leftFootIkPosition);
            float y=Mathf.Lerp(lastLeftFootPositionY,
                world2Local.y, footPositionLerpSpeed);
            targetIkPosition=transform.InverseTransformPoint(
                targetIkPosition);
            targetIkPosition.y+=y;
            lastLeftFootPositionY=y;
            targetIkPosition=transform.TransformPoint(targetIkPosition);
            Quaternion currRotation=animator.GetIKRotation(
                AvatarIKGoal.LeftFoot);
            Quaternion nextRotation=leftFootIkRotation * currRotation;
            animator.SetIKRotation(AvatarIKGoal.LeftFoot, nextRotation);
        }
        animator.SetIKPosition(AvatarIKGoal.LeftFoot, targetIkPosition);
        //应用右脚 IK 目标
        animator.SetIKPositionWeight(AvatarIKGoal.RightFoot, 1f);
        animator.SetIKRotationWeight(AvatarIKGoal.RightFoot, 1f);
        targetIkPosition=animator.GetIKPosition(
            AvatarIKGoal.RightFoot);
        if (rightFootRaycast)
        {
            Vector3 world2Local=transform.InverseTransformPoint(
                rightFootIkPosition);
            float y=Mathf.Lerp(lastRightFootPositionY,
                world2Local.y, footPositionLerpSpeed);
            targetIkPosition=transform.InverseTransformPoint(
                targetIkPosition);
            targetIkPosition.y+=y;
            lastRightFootPositionY=y;
            targetIkPosition=transform.TransformPoint(targetIkPosition);
            Quaternion currRotation=animator.GetIKRotation(
                AvatarIKGoal.RightFoot);
            Quaternion nextRotation=rightFootIkRotation * currRotation;
            animator.SetIKRotation(AvatarIKGoal.RightFoot, nextRotation);
        }
        animator.SetIKPosition(AvatarIKGoal.RightFoot, targetIkPosition);
    }
```

人物角色处于坡面时,未开启 FootIK 的效果如图 5-31(a)所示,开启 FootIK 后的效果如图 5-31(b)所示。

(a) 未开启FootIK　　　　(b) 开启FootIK

图 5-31　人物角色处于坡面

当人物角色处于台阶时,未开启 FootIK 的效果如图 5-32(a)所示,开启 FootIK 后的效果如图 5-32(b)所示。

(a) 未开启FootIK　　　　(b) 开启FootIK

图 5-32　人物角色处于台阶

当人物角色处于球面时,未开启 FootIK 的效果如图 5-33(a)所示,开启 FootIK 后的效果如图 5-33(b)所示。

5.6.3　IK 权重值曲线烘焙

在实现人物角色的动作时经常会遇到根据动画播放进度设置 IK 权重值的需求,例如 5.6.2 节中介绍的脚部 IK 功能,仅在静止状态时才可以将 IK 权重值始终设置为 1,如果人物角色处于行走、奔跑或者其他状态,则 IK 权重值只有在脚着地时才为 1,因此,通常会设置 IK 权重值曲线,在 OnAnimatorIK()方法中读取对应的参数值,以便设置 IK 权重值。

（a）未开启FootIK　　　　　（b）开启FootIK

图 5-33　人物角色处于球面

当人物角色的动作复杂多样时，手动为每个动画资产编辑动画曲线较为烦琐，因此可以考虑通过编辑器工具完成动画曲线的自动化生成，接下来介绍编辑器工具的制作过程。

1. 创建编辑器窗口

创建一个编辑器窗口类，通过 MenuItem 为打开该窗口提供菜单入口。当选中场景中某个挂载 Animator 组件的对象时，窗口中会列举出对应的动画状态机中包含的动画片段，提供 Bake 按钮，当单击该按钮时，开始为选中的动画片段烘焙动画曲线，代码如下：

```
//第 5 章/AnimationCurveBaker.cs

using System;
using System.Linq;
using UnityEngine;
using UnityEditor;
using Unity.EditorCoroutines.Editor;

public class AnimationCurveBaker : EditorWindow
{
    [MenuItem("Example/Animation Curve Baker")]
    public static void Open()
    {
        GetWindow<AnimationCurveBaker>().Show();
    }

    private Vector2 scroll;
    private int currentClipIndex;
```

```csharp
private void OnSelectionChange()
{
    currentClipIndex=0;
    Repaint();
}

private void OnGUI()
{
    GameObject selectedGo=Selection.activeGameObject;
    if (selectedGo==null)
    {
        EditorGUILayout.HelpBox("未选中任何对象", MessageType.Warning);
        return;
    }
    Animator animator=selectedGo.GetComponent<Animator>();
    if (animator==null || animator.runtimeAnimatorController==null)
    {
        EditorGUILayout.HelpBox("动画状态机不存在", MessageType.Warning);
        return;
    }
    AnimationClip[] clips=animator
        .runtimeAnimatorController.animationClips;
    if (clips.Length==0)
    {
        EditorGUILayout.HelpBox("Animator 中的动画片段数量为 0",
            MessageType.Warning);
        return;
    }
    scroll=GUILayout.BeginScrollView(scroll);
    for (int i=0; i<clips.Length; i++)
    {
        AnimationClip clip=clips[i];
        GUILayout.Label(clip.name, currentClipIndex==i
            ? "MeTransitionSelectHead" : "ProjectBrowserHeaderBgTop",
            GUILayout.Height(22f));
        if (Event.current.type==EventType.MouseDown &&
            GUILayoutUtility.GetLastRect()
                .Contains(Event.current.mousePosition))
        {
            Event.current.Use();
            currentClipIndex=i;
```

```
            }
        }
        GUILayout.EndScrollView();

        GUILayout.FlexibleSpace();
        GUILayout.BeginHorizontal();
        if (GUILayout.Button("Bake"))
        {
            //TODO:开始烘焙
        }
        GUILayout.EndHorizontal();
    }
}
```

效果如图 5-34 所示。

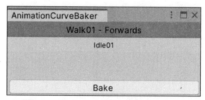

图 5-34　Animation Curve Baker

2．烘焙策略

当烘焙动画曲线时，应提供烘焙的策略，声明一个策略基类，为策略执行提供抽象方法，代码如下：

```
using UnityEngine;
using System.Collections;

public abstract class BakeStrategy
{
    public abstract IEnumerator Execute(
        Animator animator, AnimationClip clip);
}
```

可以看到，策略是在协程方法中执行的，在编辑器环境中执行协程可以使用官方提供的 Editor Coroutines 工具，该工具在 Package Manager 中导入，如图 5-35 所示。

在烘焙按钮的左侧创建一个下拉列表，用于选择烘焙策略，单击烘焙按钮后，创建对应的策略类实例，然后开启协程并根据策略烘焙动画曲线，代码如下：

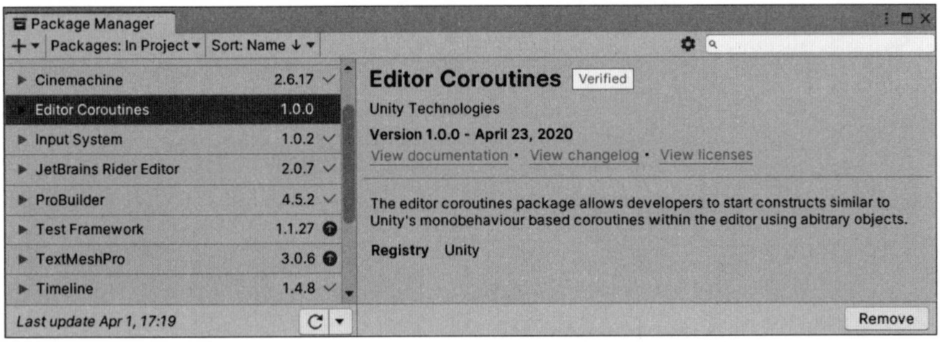

图 5-35 Editor Coroutines

```csharp
//第 5 章/AnimationCurveBaker.cs

private string[] strategyNames;
private int currentStrategyIndex;

private void OnEnable()
{
    strategyNames=typeof(BakeStrategy).Assembly
        .GetTypes()
        .Where(m =>m.IsSubclassOf(typeof(BakeStrategy)))
        .Select(m =>m.Name)
        .ToArray();
}
private void OnGUI()
{
    //...
    GUILayout.BeginHorizontal();
    currentStrategyIndex=EditorGUILayout.Popup(
        currentStrategyIndex, strategyNames);
    if (GUILayout.Button("Bake", GUILayout.Width(50f)))
    {
        Type targetType=Type.GetType(
            strategyNames[currentStrategyIndex], true);
        var strategy=Activator.CreateInstance(
            targetType) as BakeStrategy;
        EditorCoroutineUtility.StartCoroutine(
            strategy.Execute(animator, clips[currentClipIndex]), this);
    }
    GUILayout.EndHorizontal();
}
```

以脚部 IK 为例，创建一个 FootIKCurve 策略类，重写 Execute() 方法，通过 AnimationClip 类中的 SampleAnimation() 方法对人物角色进行采样，在采样之前，需要记录人物角色的初始姿态，以便在采样之后恢复其姿态，代码如下：

```
//第 5 章/BakeStrategy.cs
public class FootIKCurve : BakeStrategy
{
    public override IEnumerator Execute(
        Animator animator, AnimationClip clip)
    {
        int taskId=Progress.Start("Animation Curve Bake",
            "Baking AxisZMovement Curve", Progress.Options.None, -1);
        //获取资产路径进而获取资产导入器
        ModelImporter importer=AssetImporter.GetAtPath(
            AssetDatabase.GetAssetPath(clip)) as ModelImporter;
        //在该资产导入器中根据名称找到目标动画剪辑
        ModelImporterClipAnimation[] clipAnimations
            =importer.clipAnimations;
        ModelImporterClipAnimation target=clipAnimations
            .FirstOrDefault(m =>m.name==clip.name);
        //如果已有同名曲线,则将其筛除
        target.curves=target.curves.Where(m
            =>m.name !="LeftFootIK"
            && m.name !="RightFootIK").ToArray();
        //保存并重新导入
        importer.SaveAndReimport();
        yield return null;
        //帧数等于动画剪辑的时长乘以采样率并向上取整
        int samplingRate=30;
        int frames=Mathf.CeilToInt(clip.length * samplingRate);
        Keyframe[] lKeyframes=new Keyframe[frames];
        Keyframe[] rKeyframes=new Keyframe[frames];
        //采样之前记录初始姿态
        Dictionary<Transform, (Vector3, Quaternion)>pose
            =new Dictionary<Transform, (Vector3, Quaternion)>();
        Transform[] children=animator
            .GetComponentsInChildren<Transform>();
        for (int i=0; i<children.Length; i++)
        {
            Transform child=children[i];
```

```csharp
            pose.Add(child, (child.position, child.rotation));
            Progress.Report(taskId, (float)i / children.Length,
                "Record origin pose...");
            yield return null;
        }
        //开始采样
        for (int i=0; i<frames; i++)
        {
            //TODO:采样
            Progress.Report(taskId, (float)i / frames,
                string.Format("Sampling the frame {0}", i));
            yield return null;
        }
        //恢复初始姿态
        foreach (var kv in pose)
            kv.Key.SetPositionAndRotation(
                kv.Value.Item1, kv.Value.Item2);
        //添加曲线并保存,重新导入
        target.curves=target.curves.Concat(new ClipAnimationInfoCurve[2]
        {
            new ClipAnimationInfoCurve()
            {
                name="LeftFootIK",
                curve=new AnimationCurve(lKeyframes)
            },
            new ClipAnimationInfoCurve()
            {
                name="RightFootIK",
                curve=new AnimationCurve(rKeyframes)
            }
        }).ToArray();
        importer.clipAnimations=clipAnimations;
        importer.SaveAndReimport();
        Progress.Remove(taskId);
    }
}
```

根据脚部的坐标高度判断是否处于地面,代码如下:

```csharp
private bool IsFootGrounded(
    Animator animator, HumanBodyBones bone)
```

```
{
    Transform transform=animator.GetBoneTransform(bone);
    return transform.position.y<=.15f;
}
```

假设脚着地时权重值为 1，否则为 0，那么在采样时，可以根据 IsFootGrounded() 方法的返回值设置对应关键帧的值，代码如下：

```
//第 5 章/BakeStrategy.cs

//开始采样
for (int i=0; i<frames; i++)
{
    clip.SampleAnimation(animator.gameObject,
        (float)i / samplingRate);
    bool isFootGroundedLeft=IsFootGrounded(
        animator, HumanBodyBones.LeftFoot);
    bool isFootGroundedRight=IsFootGrounded(
        animator, HumanBodyBones.RightFoot);
    //在地面时权重值为 1,不在地面时权重值为 0
    lKeyframes[i]=new Keyframe((float)i / frames,
        isFootGroundedLeft ? 1f : 0f);
    rKeyframes[i]=new Keyframe((float)i / frames,
        isFootGroundedRight ? 1f : 0f);
    Progress.Report(taskId, (float)i / frames,
        string.Format("Sampling the frame {0}", i));
    yield return null;
}
```

需要注意的是，如果某一帧的值与其前一帧和后一帧的值均相同，则该帧是不需要的，可以过滤掉，因此，需要一种过滤的方法，代码如下：

```
//第 5 章/BakeStrategy.cs

public Keyframe[] KeyFramesFilter(Keyframe[] keyframes, int frames)
{
    keyframes=Enumerable.Range(0, frames)
        .Where(i =>
        {
            bool sameWithPrev=(i -1) >=0
                && keyframes[i -1].value
                ==keyframes[i].value;
```

```
                bool sameWithLast=(i +1)<frames
                    && keyframes[i +1].value
                    ==keyframes[i].value;
                return !sameWithPrev || !sameWithLast;
            })
            .Select(i =>keyframes[i])
            .ToArray();
        return keyframes;
    }
```

选择 FootIKCurve 策略为行走动画烘焙 IK 权重值曲线，如图 5-36 所示。烘焙完成后，可以看到行走动画多了 LeftFootIK 和 RightFootIK 两条动画曲线，如图 5-37 所示。

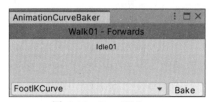

图 5-36　FootIKCurve

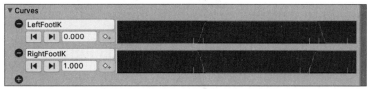

图 5-37　烘焙结果

第 6 章　寻路算法

寻路算法是游戏中人工智能的重要组成部分，它负责为游戏寻路单位规划出从起点到终点的最佳路径。随着游戏复杂度的不断提升，对于高效、灵活的寻路算法的需求也日益增长。

本章将介绍几种主流的寻路算法，包括 Unity 内置的自动导航功能 Navigation、经典的 A 星寻路算法、创新的流场寻路算法，以及基于八叉树实现的三维空间寻路算法。通过对这些算法的介绍和实现，读者将能够深入了解不同寻路算法的原理和优缺点，并学会在实际项目中灵活地应用这些算法来解决实际问题。

6.1　Navigation

Navigation 是 Unity 提供的能够让角色在场景中根据导航网格进行寻路的系统，本节将介绍导航系统工作的 4 个要素，分别是导航网格、导航网格代理、导航网格障碍物及网格外链接。

6.1.1　导航网格

为场景中的静态游戏物体设置 Navigation Static，如图 6-1 所示，通过菜单 Window/AI/Navigation 打开 Navigation 窗口，如图 6-2 所示，单击 Bake 按钮系统便会开始烘焙当前场景的导航网格。

在烘焙导航网格前调整相关烘焙设置以匹配寻路代理大小，参数详解见表 6-1。除此之外，还可以设置导航区域，系统内置了 3 种导航区域，分别为 Walkable、Not Walkable、Jump，开发者可以自定义导航区域，如图 6-3 所示。在 Object 窗口中为游戏物体指定导航区域，如图 6-4 所示。

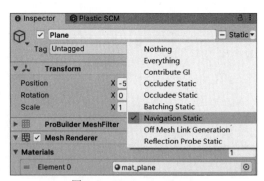

图 6-1　Navigation Static

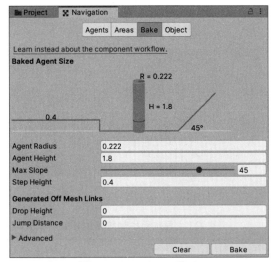

图 6-2　Navigation 窗口

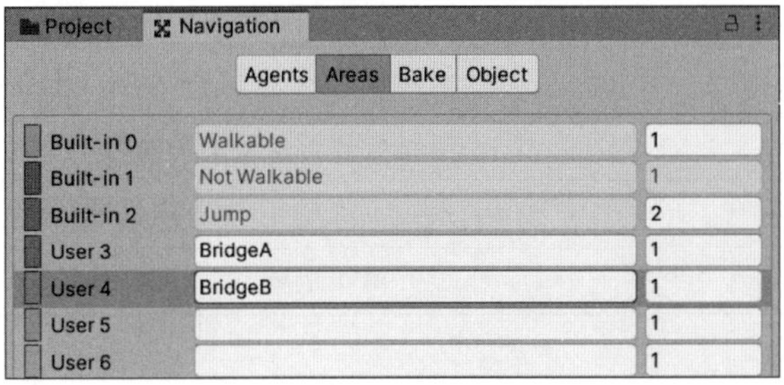

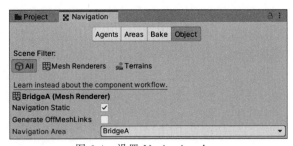

图 6-3　Navigation Areas

图 6-4　设置 Navigation Area

在烘焙过程中，系统将会收集场景中所有标记为 Navigation Static 的游戏物体的渲染网格和地形，然后处理它们以创建可行走表面的导航网格。

烘焙完成后，系统将会在当前场景所在的文件夹中创建一个新的文件夹，该场景的导航网格数据存储在这个新文件夹中，如图 6-5 所示。

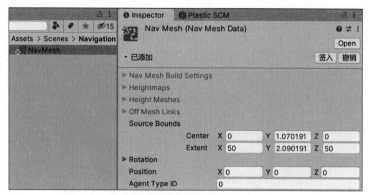

图 6-5　Nav Mesh Data

表 6-1　导航网格烘焙相关参数

参　　数	详　　解
Agent Radius	寻路代理的半径
Agent Height	寻路代理的高度
Max Slope	寻路代理可走上的最大坡面坡度
Step Height	寻路代理可走上的最大台阶高度
Drop Height	最大下落高度，用于网格外链接
Jump Distance	最大跳跃距离，用于网格外链接

6.1.2　导航网格代理

导航网格代理组件挂载于寻路单位，用于为寻路单位提供设置目的地及导航寻路功能，设置目的地的接口为 SetDestination()，需要传入一个 Vector3 类型的参数，示例代码如下：

```
//第 6 章/NavigationExample.cs

using UnityEngine;
using UnityEngine.AI;

public class NavigationExample : MonoBehaviour
{
    [SerializeField] private NavMeshAgent[] agents;
    [SerializeField] private Transform destination;

    private void OnGUI()
    {
        if (GUILayout.Button("设置寻路目的地",
```

```
            GUILayout.Width(200f), GUILayout.Height(50f)))
    {
        for (int i=0; i<agents.Length; i++)
            agents[i].SetDestination(destination.position);
    }
}
```

寻路代理寻路时的路径可以通过 path 属性获取，path 属性为 NavMeshPath 类型，该类中的 corners 属性是一个 Vector3 类型的数组，它表示的是路径的点位集合。通过 Gizmos 将路径绘制出来，代码如下：

```
//第 6 章/NavigationExample.cs
private void OnDrawGizmos()
{
    Gizmos.color=Color.cyan;
    Gizmos.DrawWireSphere(destination.position, .3f);
    Gizmos.color=Color.white;
    for (int i=0; i<agents.Length; i++)
    {
        Vector3[] path=agents[i].path.corners;
        for (int j=0; j<path.Length -1; j++)
        {
            Gizmos.DrawLine(path[j], path[j +1]);
        }
    }
}
```

结果如图 6-6 所示。

图 6-6 绘制路径

如果要判断寻路单位是否到达了目的地,则可以通过计算寻路单位的坐标与目的地之间的距离来判断距离是否小于或等于寻路代理的停止距离,代码如下:

```
//第6章/NavigationExample.cs

private void Update()
{
    for (int i=0; i<agents.Length; i++)
    {
        NavMeshAgent agent=agents[i];
        //到达目的地
        if (Vector3.Distance(agent.transform.position,
            agent.destination)<=agent.stoppingDistance)
            agent.GetComponent<Animator>().SetBool("Move", false);
        else //未到达目的地
            agent.GetComponent<Animator>().SetBool("Move", true);
    }
}
```

停止距离指的是寻路单位在距离目的地多远之后停止寻路,该值可以在检视窗口中进行调整,如图 6-7 所示。

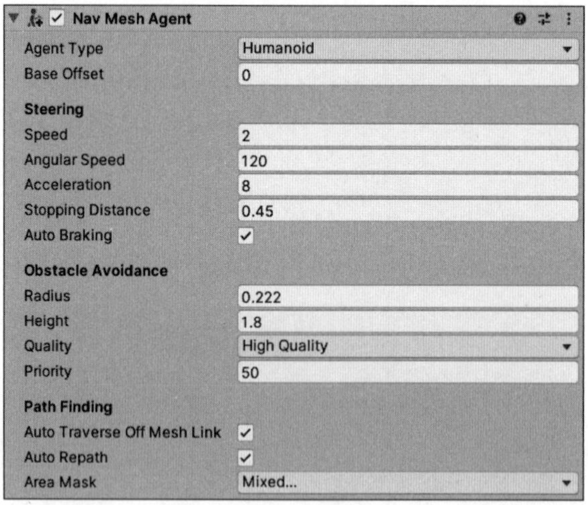

图 6-7　Nav Mesh Agent

该组件的参数详解见表 6-2。

表 6-2　Nav Mesh Agent 组件参数详解

参　　数	详　　解
Agent Type	寻路代理的类型

续表

参　　数	详　　解
Base Offset	碰撞圆柱体相对于轴心点的偏移
Speed	寻路时的移动速度
Angular Speed	寻路时的旋转速度
Acceleration	寻路时的加速度
Stopping Distance	当寻路单位与目的地的距离到达此值时,停止寻路
Auto Braking	启用后,寻路单位在到达目的地时将减速
Radius	寻路代理的半径,用于计算与其他障碍物的碰撞
Height	寻路代理的高度
Quality	避障的质量,如果代理数量较多,则可以通过降低质量来节省 CPU 开销
Priority	优先级,避障时将忽略优先级较低的代理,值越小优先级越高
Auto Traverse Off Mesh Link	是否采用默认方式经过网格外链接
Auto Repath	当现有的路径变为无效时是否尝试获取一条新路径
Area Mask	用于设置该寻路代理可以行走哪些导航区域

6.1.3　导航网格障碍物

将静态的游戏物体作为障碍物,可以通过设置 Navigation Static 使其参与导航网格烘焙,而 Nav Mesh Obstacle 组件挂载于动态的游戏物体,如图 6-8 所示。

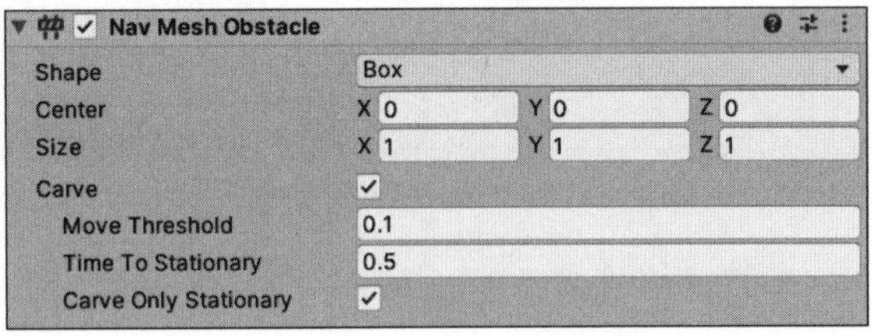

图 6-8　Nav Mesh Obstacle

寻路代理在寻路时会尽量避开它们,参数详解见表 6-3。

表 6-3　Nav Mesh Obstacle 组件参数详解

参　　数	详　　解
Shape	障碍物几何体的形状,包含盒体、胶囊体两种类型
Center(Box)	盒体的中心
Size(Box)	盒体的大小

续表

参数	详解
Center(Capsule)	胶囊体的中心
Radius(Capsule)	胶囊体的半径
Height(Capsule)	胶囊体的高度
Carve	启用时,障碍物将在导航网格中雕孔
Move Threshold	当障碍物移动的距离超过该值时将被视为非静止状态
Time To Stationary	将障碍物视为静止状态所需等待的时间,以秒为单位
Carve Only Stationary	启用后,只有障碍物为静止状态时才会在导航网格中雕孔

6.1.4 网格外链接

Off Mesh Link 组件用于合并无法使用可行走表面来表示的导航路径,如图 6-9 所示,从位置 A 到位置 B,便可以使用该组件表示。

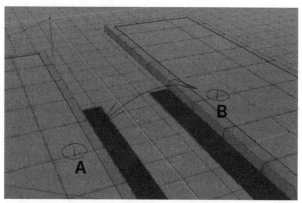

图 6-9 Off Mesh Link 使用场景

组件如图 6-10 所示,参数详解见表 6-4。

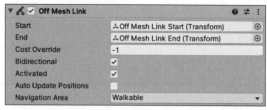

图 6-10 Off Mesh Link

表 6-4 Off Mesh Link 组件参数详解

参 数	详 解
Start	描述网格外链接起始位置的对象
End	描述网格外链接结束位置的对象
Cost Override	如果为正值,则在计算处理路径请求的路径成本时使用该值
Bi Directional	当值为 true 时表示可以在任一方向上使用链接,当值为 false 时表示只能按照从 Start 到 End 的方向使用链接
Activated	是否启用此链接
Auto Update Positions	当端点移动时是否自动更新链接
Navigation Area	描述链接的导航区域类型

6.2 A 星寻路

如果项目不涉及服务器端,则 Unity 的导航寻路功能应该能满足大部分需求,但如果涉及服务器端且使用状态同步技术,则可能需要服务器端同时实现寻路功能,这时就需要考虑其他方式实现寻路功能,而 A 星寻路算法则是常使用的一种。

A 星寻路算法是一种静态路网中求解最短路径的有效搜索方法,是启发式搜索算法中的一种,它结合了最佳优先搜索和 Dijkstra 算法的优点,能够快速地在图中找到一条从起点到终点的最短路径。

该算法主要通过不断地维护节点的代价来寻求代价最小的路径,代价的估价公式如下:

$$F(n) = G(n) + H(n) \tag{6-1}$$

其中,$G(n)$ 表示起始节点到当前节点的代价,$H(n)$ 是启发式搜索中最为关键的部分,表示当前节点到目标节点的代价。

当然,A 星算法也存在一些缺陷,例如,在游戏地图庞大、路径复杂的情况下,路径搜索过程可能需要计算成千上万个节点,计算量非常巨大,这可能降低游戏的运行速度。在实际应用中,需要根据具体需求和环境进行权衡和优化,下面将介绍 A 星算法在 Unity 中的实现过程。

6.2.1 地图数据

地图数据通过网格进行描述,如图 6-11 所示,地图网格以左下角为起点,由 $x \times y$ 个节点组成,通过 x、y 索引值获取对应的节点,每个节点对应一块场景区域,区域可能是可行走区域,也可能是障碍区域。

定义节点类,字段包含节点的 x、y 索引值,父节点信息,G、H、F 代价值及是否为可通行区域的标识信息,代码如下:

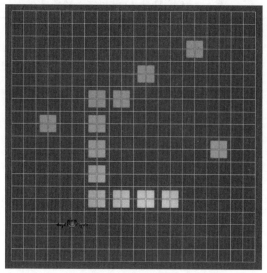

图 6-11 地图网格

```
//第6章/AStarPathFinding.GridNode.cs

public class AStarPathFinding_GridNode
{
    public int x;
    public int y;
    public AStarPathFinding_GridNode parent;
    public bool isWalkable;
    public int gCost;
    public int hCost;

    public int Cost
    {
        get
        {
            return gCost +hCost;
        }
    }

    public AStarPathFinding_GridNode(int x, int y, bool isWalkable)
    {
        this.x=x;
        this.y=y;
        this.isWalkable=isWalkable;
```

```
    }
}
```

网格类中通过字典存储所有的节点,在构造函数中进行初始化,并提供根据索引获取指定节点的方法,例如,如果 x 为 2,y 为 1,获取的则是第 2 行,第 3 列对应的节点,如图 6-12 所示。

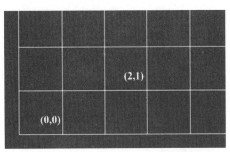

图 6-12 $x=2,y=1$ 对应的节点

在初始化时需要根据地图数据指定节点是否为可行走区域,地图数据可以通过 bool 类型的二维数组表示,也可以通过 Texture2D 类型的贴图资产表示,假设当字节为 255 时表示可行走区域,当字节为 0 时表示障碍区域,代码如下:

```
//第 6 章/AStarPathFinding.Grid.cs
using System;
using UnityEngine;
using System.Collections.Generic;
public class AStarPathFinding_Grid
{
    private readonly int x;
    private readonly int y;
    private readonly Dictionary<int, AStarPathFinding_GridNode>nodesDic;

    public AStarPathFinding_Grid(int x, int y, bool[,] map)
    {
        this.x=x;
        this.y=y;
        nodesDic=new Dictionary<int, AStarPathFinding_GridNode>();
        for (int i=0; i<x; i++)
        {
            for (int j=0; j<y; j++)
            {
```

```csharp
            int index=i * x +j;
            nodesDic.Add(index, new AStarPathFinding_GridNode(
                i, j, map[i, j]));
        }
    }
}
public AStarPathFinding_Grid(int x, int y, Texture2D map)
{
    this.x=x;
    this.y=y;
    nodesDic=new Dictionary<int, AStarPathFinding_GridNode>();
    byte[] bytes=map.GetRawTextureData();
    if (bytes.Length !=x * y)
        throw new ArgumentOutOfRangeException();
    for (int i=0; i<x; i++)
    {
        for (int j=0; j<y; j++)
        {
            int index=i * x +j;
            nodesDic.Add(index, new AStarPathFinding_GridNode(
                i, j, bytes[index]==255));
        }
    }
}
//<summary>
//根据索引获取节点
//</summary>
//<param name="x">x</param>
//<param name="y">y</param>
//<returns>节点</returns>
public AStarPathFinding_GridNode GetNode(int x, int y)
{
    return (x >=0 && x<this.x && y >=0 && y<this.y)
        ? nodesDic[x * this.y +y] : null;
}
```

6.2.2　计价方式

每向正上、正下、左、右方向走一步的代价为 1，根据勾股定理，每向斜方向走一步的代价为根号 2，近似值为 1.414，而为了便于计算、提高性能，将正方向移动一个节点的代价记

为10,将斜方向移动一个节点的代价记为14,计价函数的代码如下:

```csharp
//第 6 章/AStarPathFinding.Grid.cs

//计算两个节点之间的代价
private int CalculateCost(AStarPathFinding_GridNode node1,
    AStarPathFinding_GridNode node2)
{
    //取绝对值
    int deltaX=node1.x -node2.x;
    if (deltaX<0) deltaX=-deltaX;
    int deltaY=node1.y -node2.y;
    if (deltaY<0) deltaY=-deltaY;
    int delta=deltaX -deltaY;
    if (delta<0) delta=-delta;
    //每向上、下、左、右方向移动一个节点的代价增加 10
    //每斜方向移动一个节点的代价增加 14(勾股定理,精确来讲近似于 14.14)
    return 14 * (deltaX >deltaY ? deltaY : deltaX) +10 * delta;
}
```

6.2.3 邻节点搜索方式

搜索邻节点时有四连通和八连通两种方式,四连通指的是对应节点的上、下、左、右4个方向紧邻的节点是邻节点,如图6-13(a)所示,八连通指的是对应节点的上、下、左、右、左上、右上、左下、右下8个方向紧邻的节点是邻节点,如图6-13(b)所示,定义这两种方式,代码如下:

```csharp
//第 6 章/AStarPathFinding.SearchMode.cs

//<summary>
//邻节点搜索方式
//</summary>
public enum AStarPathFinding_SearchMode
{
    //<summary>
    //四连通
    //</summary>
    Link4,
    //<summary>
    //八连通
    //</summary>
```

```
            Link8,
}
```

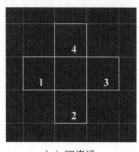 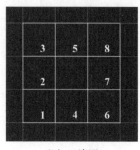

（a）四连通　　　　　　　（b）八连通

图 6-13　邻节点搜索方式

当根据四连通搜索邻节点时，只需将 x 和 y 索引值分别加 1 和减 1，便可以获得 4 个邻节点的索引。当根据八连通搜索邻节点时，可以通过双层 for 循环，依次遍历以当前节点形成的九宫格区域的节点，当前节点可能为网格的边缘节点，因此还需要考虑索引是否越界。邻节点搜索函数的代码如下：

```
//第 6 章/AStarPathFinding.Grid.cs

//获取指定节点的邻节点
public List<AStarPathFinding_GridNode>GetNeighbouringNodes(
    AStarPathFinding_GridNode node, AStarPathFinding_SearchMode searchMode)
{
    List<AStarPathFinding_GridNode>neighbours =
        new List<AStarPathFinding_GridNode>();
    switch (searchMode)
    {
        case AStarPathFinding_SearchMode.Link4:
            for (int i=-1; i<=1; i++)
            {
                if (i==0) continue;
                int x=node.x +i;
                if (x >=0 && x<this.x)
                    neighbours.Add(nodesDic[x * this.x +node.y]);
                int y=node.y +i;
                if (y >=0 && y<this.y)
                    neighbours.Add(nodesDic[node.x * this.x +y]);
            }
            break;
```

```csharp
        case AStarPathFinding_SearchMode.Link8:
            for (int i=-1; i<=1; i++)
            {
                for (int j=-1; j<=1; j++)
                {
                    if (i==0 && j==0) continue;
                    int x=node.x +i;
                    int y=node.y +j;
                    if (x >=0 && x<this.x && y >=0 && y<this.y)
                        neighbours.Add(nodesDic[x * this.x +y]);
                }
            }
            break;
    }
    return neighbours;
}
```

6.2.4 算法实现

在了解算法之前，需要先了解两个概念，即开放集合与封闭集合。开放集合是用于记录所有被考虑用来寻找最短路径的节点集合，封闭集合则是用于记录不会被考虑用来寻找最短路径的节点集合。了解了这两个概念后，下面来看算法的实现思路：

（1）将起始节点放入开放集合。

（2）开启循环，直到开放集合中节点的数量为 0，否则重复以下步骤：

① 在开放集合中寻找代价最小的节点，并把寻找到的节点作为当前节点；

② 将获取的当前节点从开放集合移除并放入封闭集合；

③ 若当前节点已经是终节点，则寻路结束，跳出循环，否则继续执行以下操作；

④ 获取当前节点的邻节点，并对每个邻节点执行以下步骤：

若邻节点是不可行走区域或者邻节点已经在封闭集合中，则不执行任何操作，继续遍历下一个邻节点；

若邻节点不在开放集合中，则将其放入开放集合，并将该邻节点的父节点设为当前节点，计算、记录该邻节点的 G、H 代价；

若邻节点在开放集合中，则判断经当前节点到达该邻节点的 G 值是否小于原来的 G 值，若小于原来的 G 值，则将该邻节点的父节点设为当前节点，并重新计算该邻节点的 G、H 代价。

（3）从终节点开始依次获取父节点并放入一个列表，最终对列表进行倒序操作就是最终寻路的路径。

代码如下：

```csharp
//第 6 章/AStarPathFinding.Grid.cs

//<summary>
//根据起始节点和终节点获取路径
//</summary>
//<param name="startNode">起始节点</param>
//<param name="endNode">终节点</param>
//<param name="searchMode">邻节点的搜索方式 四连通/八连通</param>
//<returns>路径节点集合</returns>
public List<AStarPathFinding_GridNode> GetPath(
    AStarPathFinding_GridNode startNode,
    AStarPathFinding_GridNode endNode,
    AStarPathFinding_SearchMode searchMode)
{
    if (!endNode.isWalkable) return null;
    //开放集合
    List<AStarPathFinding_GridNode> openCollection
        = new List<AStarPathFinding_GridNode>();
    //封闭集合
    List<AStarPathFinding_GridNode> closeCollection
        = new List<AStarPathFinding_GridNode>();
    //将起始节点放入开放集合
    openCollection.Add(startNode);
    //当开放集合中节点的数量为 0 时寻路结束
    while (openCollection.Count > 0)
    {
        //当前节点
        AStarPathFinding_GridNode currentNode = openCollection[0];
        //遍历查找是否有代价更小的节点
        //若代价相同,则选择移动到终点代价更小的节点
        for (int i = 1; i < openCollection.Count; i++)
        {
            currentNode = (currentNode.Cost > openCollection[i].Cost
                || (currentNode.Cost == openCollection[i].Cost
                    && currentNode.hCost > openCollection[i].hCost))
                ? openCollection[i] : currentNode;
        }
        //将获取的当前节点从开放集合移除并放入封闭集合
        openCollection.Remove(currentNode);
        closeCollection.Add(currentNode);
        //当前节点已经是终节点,寻路结束
```

```csharp
            if (currentNode==endNode)
                break;
            //获取邻节点
            List<AStarPathFinding_GridNode>neighbourNodes
                =GetNeighbouringNodes(currentNode, searchMode);
            //在当前节点向邻节点继续搜索
            for (int i=0; i<neighbourNodes.Count; i++)
            {
                AStarPathFinding_GridNode neighbourNode=neighbourNodes[i];
                //判断邻节点是否为不可行走区域(障碍)或者邻节点已经在封闭集合中
                if (!neighbourNode.isWalkable
                    || closeCollection.Contains(neighbourNode))
                    continue;

                //经当前节点到达该邻节点的G值是否小于原来的G值
                //或者该邻节点还没有放入开放集合,将其放入开放集合
                int gCost=currentNode.gCost
                    +CalculateCost(currentNode, neighbourNode);
                if (gCost<neighbourNode.gCost
                    || !openCollection.Contains(neighbourNode))
                {
                    neighbourNode.gCost=gCost;
                    neighbourNode.hCost=CalculateCost(neighbourNode, endNode);
                    neighbourNode.parent=currentNode;
                    if (!openCollection.Contains(neighbourNode))
                        openCollection.Add(neighbourNode);
                }
            }
        }
        //倒序获取父节点
        List<AStarPathFinding_GridNode>path
            =new List<AStarPathFinding_GridNode>();
        AStarPathFinding_GridNode currNode=endNode;
        while (currNode!=startNode)
        {
            path.Add(currNode);
            currNode=currNode.parent;
        }
        //再次倒序后得到完整路径
        path.Reverse();
        return path;
    }
}
```

6.2.5 寻路组件

当设置寻路目的地时,参数是一个坐标值,因此寻路组件中需要提供根据坐标值获取对应节点的方法,这与网格节点的大小相关。将坐标各维度值除以节点大小并向下取整得到索引值,根据索引值获取节点。为了方便判断寻路单位是否到了目标节点,寻路组件还需要提供根据节点获取对应坐标值的方法,该坐标值指的是节点的中心点坐标,代码如下:

```
//第 6 章/AStarPathFinding.cs

using UnityEngine;
using System.Collections.Generic;

public class AStarPathFinding : MonoBehaviour
{
    //x、y 形成地图节点网格
    [SerializeField] private int x=10;
    [SerializeField] private int y=10;
    //节点的大小
    [SerializeField] private float nodeSize=1f;
    //起点偏移值,默认以原点为网格起点
    [SerializeField] private Vector3 offset;
    //存储可行走区域数据的地图
    [SerializeField] private Texture2D map;
    private AStarPathFinding_Grid grid;

    private void Start()
    {
        grid=new AStarPathFinding_Grid(x, y, map);
    }
    //<summary>
    //根据坐标值获取节点
    //</summary>
    //<param name="position">坐标</param>
    //<returns>节点</returns>
    public AStarPathFinding_GridNode GetClosestNode(Vector3 position)
    {
        return grid?.GetNode(
            Mathf.FloorToInt((position.x -offset.x) / nodeSize),
            Mathf.FloorToInt((position.z-offset.z) / nodeSize));
    }    //<summary>
    //根据节点获取坐标值(节点中心坐标)
```

```
//</summary>
//<param name="node">节点</param>
//<returns>节点对应的坐标点</returns>
public Vector3 GetPosition(AStarPathFinding_GridNode node)
{
    return new Vector3(
        (node.x + .5f) * nodeSize,
        0f,
        (node.y + .5f) * nodeSize) +offset;
}
//<summary>
//根据起始节点和终节点获取路径
//</summary>
//<param name="startNode">起始节点</param>
//<param name="endNode">终节点</param>
//<param name="searchMode">邻节点搜索方式</param>
//<returns>路径节点集合</returns>
public List<AStarPathFinding_GridNode>GetPath(
    AStarPathFinding_GridNode startNode,
    AStarPathFinding_GridNode endNode,
    AStarPathFinding_SearchMode searchMode
        =AStarPathFinding_SearchMode.Link8)
{
    return (startNode !=null && endNode !=null)
        ? grid?.GetPath(startNode, endNode, searchMode)
        : null;
}
}
```

6.2.6 寻路代理

AStarPathFinding 组件用于承载地图数据,根据地图数据执行算法,提供最短路径,在场景中是全局唯一的。提供设置寻路目的地接口并具体实现沿路径移动的是寻路代理组件,该组件挂载于每个寻路单位,代码如下:

```
//第 6 章/AStarPathFindingAgent.cs

using UnityEngine;
using System.Collections.Generic;

public class AStarPathFindingAgent : MonoBehaviour
```

```csharp
{
    [SerializeField] private AStarPathFinding aStar;
    [SerializeField] private float moveSpeed=2f;
    [SerializeField] private float rotateSpeed=10f;
    [Tooltip("应小于网格节点大小")]
    [SerializeField] private float stopDistance=.1f;
    private List<AStarPathFinding_GridNode>path;
    private bool isStopped=true;

    public bool IsStopped
    {
        get { return isStopped; }
        set
        {
            isStopped=value;
            if (!isStopped)
            {
                path.Clear();
                path=null;
            }
        }
    }
    public Vector3 Destination { get; private set; }

    //<summary>
    //设置寻路目的地
    //</summary>
    //<param name="destination">目的地</param>
    public void SetDestination(Vector3 destination)
    {
        var targetNode=aStar.GetClosestNode(destination);
        if (targetNode.isWalkable)
        {
            isStopped=false;
            Destination=destination;
            var startNode=aStar.GetClosestNode(transform.position);
            path=aStar.GetPath(startNode, targetNode);
        }
    }
    private void Update()
    {
```

```csharp
        if (!isStopped && path !=null && path.Count >0)
        {
            var node=path[0];
            //将最终节点移动至目的地
            //将非最终节点移动至节点中心
            Vector3 des=path.Count !=1
                ? aStar.GetPosition(node)
                : Destination;
            //与目的地的距离大于停止距离
            if (Vector3.Distance(transform.position, des) >stopDistance)
            {
                //移动方向
                Vector3 direction=(des -transform.position).normalized;
                transform.position+=Time.deltaTime * moveSpeed * direction;
                transform.rotation=Quaternion.Lerp(transform.rotation,
                    Quaternion.LookRotation(direction),
                    Time.deltaTime * rotateSpeed);
            }
            else
            {
                path.RemoveAt(0);
                isStopped=path.Count==0;
            }
        }
    }

    private void OnDrawGizmosSelected()
    {
#if UNITY_EDITOR
        if (path !=null && path.Count >1)
        {
            for (int i=0; i<path.Count -1; i++)
            {
                UnityEditor.Handles.DrawLine(aStar.GetPosition(path[i]),
                    aStar.GetPosition(path[i +1]), 3f);
            }
            UnityEditor.Handles.Label(Destination,
                "目的地", new GUIStyle(GUI.skin.label)
                {
                    fontSize=20,
                    fontStyle=FontStyle.Bold
```

```
            });
#endif
        }
    }
}
```

实现一个使用寻路代理进行寻路的示例，单击后，在鼠标位置进行射线投射检测，将碰撞的位置设为寻路目的地，代码如下：

```
//第 6 章/AStarPathFindingExample.cs
using UnityEngine;

public class AStarPathFindingExample : MonoBehaviour
{
    private Animator animator;
    private AStarPathFindingAgent agent;
    private void Start()
    {
        animator=GetComponent<Animator>();
        agent=GetComponent<AStarPathFindingAgent>();
    }
    private void Update()
    {
        if (Input.GetMouseButtonDown(0))
        {
            Ray ray=Camera.main.ScreenPointToRay(Input.mousePosition);
            if (Physics.Raycast(ray, out RaycastHit hitInfo))
                agent.SetDestination(hitInfo.point);
        }
        animator.SetBool("Move", !agent.IsStopped);
    }
}
```

运行结果如图 6-14 所示。

6.2.7 路径优化

读者看到此处可能会感到疑惑，通过算法得出的路径不是已经是最优解了吗？为什么还要优化？这是由于地图是由网格节点组成的，网格节点越大，问题越明显，仔细观察图 6-14 可以发现，从出发地到目的地明显有更为平滑的路径，如图 6-15 所示。

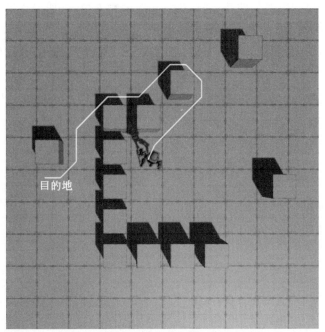

图 6-14 A 星寻路示例

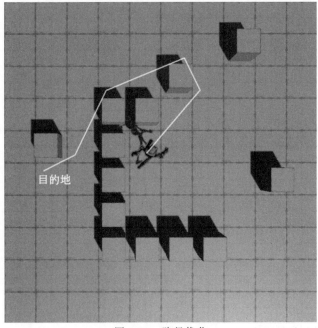

图 6-15 路径优化

为了得到这条更为平滑的路径，需要进行相关的物理检测，如射线投射检测或者球体投射检测，视具体情况而定。此处以球体投射检测为例，将路径的第 1 个节点作为当前节点，从当前节点依次向路径中的后续节点进行球体投射检测，直到检测到碰撞器时，才将上一个没有检测到碰撞器的节点放进新路径，并将其作为当前节点，代码如下：

```
//第 6 章/AStarPathFinding.cs

//<summary>
//路径平滑
//</summary>
//<param name="path">原始路径</param>
//<returns>更为平滑的路径</returns>
public List<AStarPathFinding_GridNode>Smooth(
    List<AStarPathFinding_GridNode>path)
{
    if (path.Count<=2) return path;
    List<AStarPathFinding_GridNode>smoothPath
        =new List<AStarPathFinding_GridNode>();
    int i, j;
    for (i=0; i<path.Count -1; i++)
    {
        for (j=i +1; j<path.Count; j++)
        {
            var currNode=path[i];
            var nextNode=path[j];
            Vector3 currPos=GetPosition(currNode);
            Vector3 nextPos=GetPosition(nextNode);
            Ray ray=new Ray(currPos, nextPos -currPos);
            if (Physics.SphereCast(ray, 0.15f,
                (nextPos -currPos).magnitude))
                break;
        }
        i=j -1;
        smoothPath.Add(path[i]);
    }
    return smoothPath;
}
```

6.2.8 地图编辑器

通过 EditorWindow 实现一个地图编辑器窗口工具，如图 6-16 所示，当该窗口打开时，为SceneView.duringSceneGui添加委托事件,在委托事件中使用Handles类中的DrawLine()方法绘制网格。使用Ctrl+鼠标左键快速编辑障碍区域的网格节点,相应地,使用Alt+鼠标左键快速编辑可行走区域的网格节点,它们依赖于射线投射检测实现。

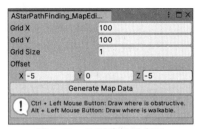

图6-16 地图编辑器窗口

地图数据编辑完成后,创建一个Texture2D类型的贴图资产,用于存储地图数据,贴图的宽和高对应地图网格的 x 与 y,贴图的字节数据根据地图数据进行填充,当字节为255时表示可行走区域,当字节为0时表示障碍区域,代码如下:

```
//第6章/AStarPathFinding.MapEditorWindow.cs

using UnityEngine;
using UnityEditor;

public class AStarPathFinding_MapEditorWindow : EditorWindow
{
    [MenuItem("AStar/Map Editor")]
    public static void Open()
    {
        GetWindow<AStarPathFinding_MapEditorWindow>().Show();
    }

    private int x=10;
    private int y=10;
    private float nodeSize=1f;
    private Vector3 offset;
    private bool[,] map;

    private void OnEnable()
    {
        SceneView.duringSceneGui -=OnSceneGUI;
        SceneView.duringSceneGui+=OnSceneGUI;
        OnMapChanged();
    }
    private void OnGUI()
    {
        EditorGUI.BeginChangeCheck();
        x=EditorGUILayout.IntField("Grid X", x);
        y=EditorGUILayout.IntField("Grid Y", y);
```

```csharp
nodeSize=EditorGUILayout.FloatField("Grid Size", nodeSize);
offset=EditorGUILayout.Vector3Field("Offset", offset);
if (EditorGUI.EndChangeCheck())
{
    OnMapChanged();
    SceneView.RepaintAll();
}
//生成地图
if (GUILayout.Button("Generate Map Data"))
{
    //选择保存路径
    string filePath=EditorUtility.SaveFilePanel(
        "Save Map Data", Application.dataPath,
        "New Map Data", "asset");
    if (!string.IsNullOrEmpty(filePath))
    {
        //转换为 Asset 路径
        filePath=filePath.Substring(
            filePath.IndexOf("Assets"));
        //创建地图 Texture
        Texture2D bitmap=new Texture2D(x, y,
            TextureFormat.Alpha8, false);
        byte[] bytes=bitmap.GetRawTextureData();
        for (int m=0; m<x; m++)
        {
            for (int n=0; n<y; n++)
            {
                //0 表示障碍区域,255 表示可行走区域
                bytes[m * x +n]=(byte)(map[m, n] ? 255 : 0);
            }
        }
        bitmap.LoadRawTextureData(bytes);
        //创建、保存资产
        AssetDatabase.CreateAsset(bitmap, filePath);
        AssetDatabase.SaveAssets();
        AssetDatabase.Refresh();
        //选中
        EditorGUIUtility.PingObject(bitmap);
    }
}
EditorGUILayout.HelpBox(
```

```csharp
            "Ctrl +Left Mouse Button: Draw where is obstructive." +
            "\r\nAlt +Left Mouse Button: Draw where is walkable.",
            MessageType.Info);
}
private void OnSceneGUI(SceneView sceneView)
{
    //绘制地图网格
    Handles.color=Color.cyan;
    for (int i=0; i<=x; i++)
    {
        Vector3 start=i * nodeSize * Vector3.right;
        Vector3 end=start +y * nodeSize * Vector3.forward;
        Handles.DrawLine(start +offset, end +offset);
    }
    for (int i=0; i<=y; i++)
    {
        Vector3 start=i * nodeSize * Vector3.forward;
        Vector3 end=start +x * nodeSize * Vector3.right;
        Handles.DrawLine(start +offset, end +offset);
    }
    HandleUtility.AddDefaultControl(
        GUIUtility.GetControlID(FocusType.Passive));
    //Ctrl +鼠标左键 绘制障碍区域
    //Alt +鼠标左键 绘制可行走区域
    var e=Event.current;
    if (e !=null && (e.control || e.alt)
        && (e.type==EventType.MouseDown
            || e.type==EventType.MouseDrag)
        && e.button==0)
    {
        Ray ray=HandleUtility.GUIPointToWorldRay(e.mousePosition);
        if (Physics.Raycast(ray, out RaycastHit hit))
        {
            int targetX=Mathf.CeilToInt(
                (hit.point.x -offset.x) / nodeSize);
            int targetY=Mathf.CeilToInt(
                (hit.point.z -offset.z) / nodeSize);
            if (targetX<=x && targetX >0 &&
                targetY<=y && targetY >0)
                map[targetX -1, targetY -1]=!e.control;
        }
```

```
            e.Use();
        }
        //绘制障碍区域
        Handles.color=Color.red;
        for (int m=0; m<x; m++)
        {
            for (int n=0; n<y; n++)
            {
                if (!map[m, n])
                    Handles.DrawWireCube(new Vector3(m * nodeSize, 0f,
                        n * nodeSize) +.5f * nodeSize *
                        (Vector3.forward +Vector3.right) +offset,
                        .9f * nodeSize * (Vector3.forward +Vector3.right));
            }
        }
    }
    private void OnMapChanged()
    {
        map=new bool[x, y];
        for (int i=0; i<x; i++)
        {
            for (int j=0; j<y; j++)
            {
                map[i, j]=true;
            }
        }
    }
    private void OnDisable()
    {
        SceneView.duringSceneGui -=OnSceneGUI;
    }
}
```

6.3　流场寻路

当场景中有数十、数百，甚至数千个寻路单位需要到达同一目的地时，通过常规的 A 星算法进行寻路不再是合适的选择，因为每个寻路单位都需要依据自身所在的位置执行算法，从而获得一条到达目的地的路径，n 个游戏单位进行寻路就需要执行 n 次算法，这是比较大的性能开销，而流场寻路方法可以很恰当地解决这一问题，流场寻路算法通常用于 RTS 类型游戏中的群体寻路。

6.3.1 流场

将场景分割为 $x \times y$ 个网格节点,当确认目的地时,流场寻路算法会依据目的地所在的网格,依次向周围获取邻节点,计算邻节点到当前节点的代价。代价大的节点指向代价小的节点便可以形成一个方向,而这些节点形成的方向便组成了流场,如图 6-17 所示。

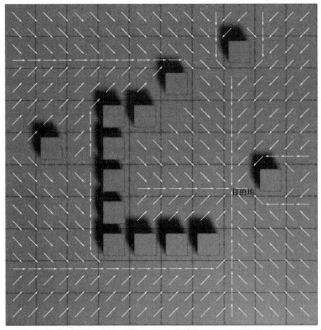

图 6-17 流场

流场形成后,寻路单位只需根据自身所在的坐标位置,获得对应的网格节点,依据节点的方向进行移动,最终便可以到达目的地。

节点类除节点索引对应的字段 x、y 之外,还包括表示代价的字段 cost 与 fCost,以及形成流场的节点方向字段。cost 为基础代价,场景中可能会有多种地形,例如水泥地、沼泽地等,不同地形有不同的代价,假设所有节点所在的地形是相同的,与 6.2.2 节中的网格节点一致,只考虑是否为可行走区域,通过字段 isWalkable 表示。fCost 是指节点到目标节点的最终代价,代码如下:

```
//第6章/FlowFieldPathFinding.GridNode.cs

using UnityEngine;

public class FlowFieldPathFinding_GridNode
{
```

```csharp
        public int x;
        public int y;
        public int cost;
        public int fCost;
        public Vector3 direction;
        public bool isWalkable;

        public FlowFieldPathFinding_GridNode(int x, int y, bool isWalkable)
        {
            this.x=x;
            this.y=y;
            this.isWalkable=isWalkable;
            cost=isWalkable ? 10 : int.MaxValue;
            fCost=int.MaxValue;
        }
    }
```

6.3.2 算法实现

6.2.2 节中介绍了节点的计价方式、搜索邻节点的方法等相关内容，此处不再介绍。流场寻路的网格类中需要提供设置目标节点的方法，当设置目标节点时，将其放入一个队列中，当队列的数量大于 0 时，不断地执行以下步骤。

（1）出列，获得当前节点。
（2）搜索当前节点的邻节点。
（3）遍历当前节点的邻节点，计算邻节点到当前节点的基础代价。
（4）将邻节点到当前节点的基础代价加上当前节点的最终代价记为 t，如果 t 小于邻节点的最终代价，则将邻节点的最终代价设置为 t，并将其放入队列。

代码如下：

```csharp
//第 6 章/FlowFieldPathFinding.Grid.cs

    //<summary>
//设置目标节点
//</summary>
//<param name="target">目标节点</param>
public void SetTarget(FlowFieldPathFinding_GridNode target)
{
    foreach (var node in nodesDic.Values)
    {
        node.cost=node.isWalkable ? 10 : int.MaxValue;
```

```
            node.fCost=int.MaxValue;
        }
        target.cost=0;
        target.fCost=0;
        target.direction=Vector3.zero;
        Queue<FlowFieldPathFinding_GridNode>queue
            =new Queue<FlowFieldPathFinding_GridNode>();
        queue.Enqueue(target);
        while (queue.Count >0)
        {
            FlowFieldPathFinding_GridNode currentNode=queue.Dequeue();
            List<FlowFieldPathFinding_GridNode>neighbourNodes =
                GetNeighbouringNodes(currentNode,
                    FlowFieldPathFinding_SearchMode.Link8);
            for (int i=0; i<neighbourNodes.Count; i++)
            {
                FlowFieldPathFinding_GridNode neighbourNode=neighbourNodes[i];
                if (neighbourNode.cost==int.MaxValue) continue;
                neighbourNode.cost=CalculateCost(neighbourNode, currentNode);
                if (neighbourNode.cost +currentNode.fCost<neighbourNode.fCost)
                {
                    neighbourNode.fCost=neighbourNode.cost +currentNode.fCost;
                    queue.Enqueue(neighbourNode);
                }
            }
        }
    }
```

设置目标节点计算各节点的代价后便可以生成流场,生成流场时遍历节点的邻节点,如果邻节点的最终代价小于当前节点的最终代价,则当前节点的方向便是当前节点指向邻节点的方向。若出现代价相同的情况,则需要考虑哪一个邻节点更加接近目标节点,代码如下:

```
//第 6 章/FlowFieldPathFinding.Grid.cs

//<summary>
//生成流场
//</summary>
//<param name="target">目标节点</param>
public void GenerateFlowField(FlowFieldPathFinding_GridNode target)
{
```

```csharp
            foreach (var node in nodesDic.Values)
            {
                List<FlowFieldPathFinding_GridNode>neighbourNodes =
                    GetNeighbouringNodes(node,
                        FlowFieldPathFinding_SearchMode.Link8);
                int fCost=node.fCost;
                FlowFieldPathFinding_GridNode temp=null;
                for (int i=0; i<neighbourNodes.Count; i++)
                {
                    FlowFieldPathFinding_GridNode neighbourNode=neighbourNodes[i];
                    if (neighbourNode.fCost<fCost)
                    {
                        temp=neighbourNode;
                        fCost=neighbourNode.fCost;
                        node.direction=new Vector3(
                            neighbourNode.x -node.x, 0, neighbourNode.y -node.y);
                    }
                    else if (neighbourNode.fCost==fCost && temp !=null)
                    {
                        if (CalculateCost(neighbourNode, target) <
                            CalculateCost(temp, target))
                        {
                            temp=neighbourNode;
                            fCost=neighbourNode.fCost;
                            node.direction=new Vector3(
                                neighbourNode.x -node.x, 0, neighbourNode.y -node.y);
                        }
                    }
                }
            }
        }
```

6.3.3 寻路组件

流场寻路组件与 A 星寻路组件一样,包含根据节点获取坐标的方法和根据坐标获取节点的方法,除此之外,流场寻路组件提供了设置目的地的接口,设置目的地时生成流场,代码如下:

```csharp
//第 6 章/FlowFieldPathFinding.cs

using UnityEngine;
```

```csharp
public class FlowFieldPathFinding : MonoBehaviour
{
    [SerializeField] private int x=10;
    [SerializeField] private int y=10;
    [SerializeField] private float nodeSize=1f;
    //起点偏移值,默认以原点为网格起点
    [SerializeField] private Vector3 offset;
    //存储可行走区域数据的地图
    [SerializeField] private Texture2D map;
    private FlowFieldPathFinding_Grid grid;
    public FlowFieldPathFinding_GridNode TargetNode { get; private set; }
    private void Start()
    {
        grid=new FlowFieldPathFinding_Grid(x, y, map);
    }
    //<summary>
    //根据坐标值获取节点
    //</summary>
    //<param name="position">坐标</param>
    //<returns>节点</returns>
    public FlowFieldPathFinding_GridNode GetClosestNode(Vector3 position)
    {
        return grid?.GetNode(
            Mathf.FloorToInt((position.x - offset.x) / nodeSize),
            Mathf.FloorToInt((position.z - offset.z) / nodeSize));
    }
    //<summary>
    //根据节点获取坐标值(节点中心坐标)
    //</summary>
    //<param name="node">节点</param>
    //<returns>节点对应的坐标点</returns>
    public Vector3 GetPosition(FlowFieldPathFinding_GridNode node)
    {
        return new Vector3(
            (node.x + .5f) * nodeSize,
            0f,
            (node.y + .5f) * nodeSize) +offset;
    }
    //<summary>
    //设置目的地
```

```
//</summary>
//<param name="destination"></param>
public void SetDestination(Vector3 destination)
{
    var node=GetClosestNode(destination);
    if (node.isWalkable)
    {
        TargetNode=node;
        grid.SetTarget(node);
        grid.GenerateFlowField(node);
    }
}
```

在OnDrawGizmos()方法中将网格、节点的代价及流场通过Gizmos和Handles类中的方法绘制出来,实现可视化,代码如下:

```
//第6章/FlowFieldPathFinding.cs

#if UNITY_EDITOR
[SerializeField] private bool drawGrid=true;
[SerializeField] private bool drawCost=true;
[SerializeField] private bool drawFlowField=true;
private void OnDrawGizmos()
{
    if (drawGrid)
    {
        Gizmos.color=Color.cyan;
        for (int i=0; i<=x; i++)
        {
            Vector3 start=i * nodeSize * Vector3.right;
            Vector3 end=start +y * nodeSize * Vector3.forward;
            Gizmos.DrawLine(start +offset, end +offset);
        }
        for (int i=0; i<=y; i++)
        {
            Vector3 start=i * nodeSize * Vector3.forward;
            Vector3 end=start +x * nodeSize * Vector3.right;
            Gizmos.DrawLine(start +offset, end +offset);
        }
    }
```

```csharp
            if (grid !=null)
            {
                for (int i=0; i<x; i++)
                {
                    for (int j=0; j<y; j++)
                    {
                        var node=grid.GetNode(i, j);
                        if (node.isWalkable)
                        {
                            Vector3 pos=GetPosition(node);
                            if (drawCost)
                                Handles.Label(pos, node.fCost.ToString());
                            //Handles.Label(pos, node.direction.ToString());
                            if (drawFlowField)
                            {
                                if (node.direction !=Vector3.zero)
                                {
                                    Handles.ArrowHandleCap(
                                        0,
                                        pos,
                                        Quaternion.LookRotation(node.direction),
                                        nodeSize * .75f,
                                        EventType.Repaint);
                                }
                            }
                        }
                    }
                }
            }
            if (TargetNode !=null)
                Handles.Label(GetPosition(TargetNode), "目的地");
        }
#endif
```

6.3.4 寻路代理

寻路代理根据自身所在位置获取对应节点,根据节点的方向进行寻路即可,当节点的方向变为 0 向量时,表示到达了目的地,代码如下:

```
//第 6 章/FlowFieldPathFindingAgent.cs
```

```
using UnityEngine;
public class FlowFieldPathFindingAgent : MonoBehaviour
{
    [SerializeField] private FlowFieldPathFinding flowField;
    [SerializeField] private float moveSpeed=2f;
    [SerializeField] private float rotateSpeed=10f;

    public bool isPathFinding;

    private void Update()
    {
        if (flowField !=null && flowField.TargetNode !=null)
        {
            var node=flowField.GetClosestNode(transform.position);
            if (node.direction !=Vector3.zero)
            {
                isPathFinding=true;
                transform.position+=Time.deltaTime
                    * moveSpeed * node.direction;
                transform.rotation=Quaternion.Lerp(
                    transform.rotation,
                    Quaternion.LookRotation(node.direction),
                    Time.deltaTime * rotateSpeed);
            }
            else isPathFinding=false;
        }
    }
}
```

6.4 八叉树寻路

八叉树是用于描述三维空间的树状数据结构，可以将其看作四叉树在三维空间上的扩展。八叉树的节点通过一个立方体区域进行表示，如果节点对应的立方体区域内没有任何物体，则将该节点作为八叉树的叶节点，否则该立方体区域将被继续划分为8个等分的子区域，也就是说该节点包含8个子节点。通过递归的方式不断地细分节点区域，直至区域内不再有任何物体或者节点大小已经达到了指定的最小规模。

当定义节点类时，需要声明表示节点中心和节点大小的字段，以及用于存储子节点的数组，代码如下：

```
//第 6 章/OctreeNode.cs

using UnityEngine;
using System.Collections.Generic;

public class OctreeNode
{
    public Vector3 center; //节点中心
    public float nodeSize; //节点大小
    public OctreeNode[] subNodes; //子节点
    private readonly float minNodeSize;
}
```

在构造函数中,通过盒体重叠检测方法检测节点区域内是否有碰撞器,如果有,则需要继续划分子区域,子节点的大小等于当前节点大小的 1/2,子节点的中心坐标会在各个坐标轴上偏移当前节点大小的 1/4。如果当前节点大小的 1/2 小于指定的最小节点大小,则不再继续划分,代码如下:

```
//第 6 章/OctreeNode.cs

public bool isPassable;
public OctreeNode(Vector3 center, float nodeSize, float minNodeSize)
{
    this.center=center;
    this.nodeSize=nodeSize;
    this.minNodeSize=minNodeSize;

    //子节点大小等于当前节点大小的 1/2
    float subNodeSize=nodeSize * .5f;
    //盒体重叠检测,如果未检测到任何碰撞器,则表示该节点是可通行的
    isPassable=Physics.OverlapBox(center,
        nodeSize * .5f * Vector3.one).Length==0;
    //如果该区域内没有任何碰撞器,则不需要继续划分
    //如果有碰撞,并且子节点大小大于最小节点大小,则继续划分
    if (!isPassable && subNodeSize >minNodeSize)
    {
        //子节点的中心坐标会在各坐标轴上偏移当前节点大小的 1/4
        float quarter=nodeSize * .25f;
        subNodes=new OctreeNode[8];
        int index=-1;
        for (int x=-1; x<=1; x+=2)
```

```
            {
                for (int y=-1; y<=1; y+=2)
                {
                    for (int z=-1; z<=1; z+=2)
                    {
                        subNodes[++index]=new OctreeNode(
                            center +quarter * new Vector3(x, y, z),
                            subNodeSize, minNodeSize);
                    }
                }
            }
        }
    }
}
```

在八叉树的类构造函数中创建八叉树的根节点,通过递归完成空间区域的划分,代码如下:

```
//第6章/Octree.cs

using UnityEngine;
using System.Collections.Generic;

public class Octree
{
    //根节点
    public OctreeNode rootNode;

    public Octree(Vector3 center, float rootNodeSize, float minLeafSize)
    {
        rootNode=new OctreeNode(center, rootNodeSize, minLeafSize);
    }
}
```

为了便于在场景中观察空间区域划分的情况,在节点类中定义一个绘制方法,通过Gizmos类中的DrawWireCube()方法绘制节点对应的立方体区域,代码如下:

```
//第6章/OctreeNode.cs

public void Draw(bool drawCenter, bool drawConnections)
{
    if (subNodes !=null)
    {
```

```
            for (int i=0; i<subNodes.Length; i++)
            {
                subNodes[i].Draw(drawCenter, drawConnections);
            }
        }
        else
            Gizmos.DrawWireCube(center, nodeSize * Vector3.one);
}
```

在示例场景中摆放一些带有 BoxCollider 组件的 Cube 物体, 在示例脚本中创建八叉树, 效果如图 6-18 所示, 可以看到, 碰撞器所在的区域划分更加密集。

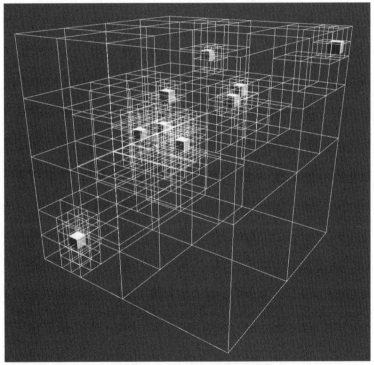

图 6-18　八叉树空间区域划分

通过八叉树数据结构, 可以实现可飞行游戏单位在三维空间中进行寻路的功能。6.2 节中讲解了 A 星寻路的算法, 将其应用到此处, 需要进行修改。

一是节点间的计价方式, 由于节点的大小是不固定的, 所以不同节点间的距离也是不同的, 因此可以直接通过计算两个节点中心点之间的距离进行计价, 代码如下:

```
//节点间计价
private float CalculateCost(OctreeNode node1, OctreeNode node2)
```

```
{
    return (node2.center -node1.center).magnitude;
}
```

二是邻节点的搜索方式,此处通过与周围的节点构建链接的方式获取邻节点。在构建节点间的链接之前,需要先获取所有的叶节点,也就是没有子节点的节点,通过递归的方式获取,代码如下:

```
//第 6 章/Octree.cs

//叶节点集合
private readonly List<OctreeNode>leavesNodes;
//获取叶节点
private void GetLeavesNodes(OctreeNode node)
{
    //该节点没有子节点,那么它就是叶节点
    if (node.subNodes==null)
        leavesNodes.Add(node);
    else
    {
        //遍历子节点
        for (int i=0; i<node.subNodes.Length; i++)
        {
            //递归获取叶节点
            GetLeavesNodes(node.subNodes[i]);
        }
    }
}
```

在节点类中,通过列表存储链接节点,在与其他节点构建链接时,需要两个节点都是可通行的节点,并且两个节点之间没有碰撞器阻挡,代码如下:

```
//第 6 章/OctreeNode.cs

public List<OctreeNode>connections; //链接节点集合

//构建与指定节点的链接
public void BuildConnection(OctreeNode node)
{
    connections ??=new List<OctreeNode>();
    if (node==null) return;
    if (connections.Contains(node)) return;
```

```
        if (!isPassable || !node.isPassable) return;
        //通过球体投射检测方法,检测两个节点间是否有碰撞器
        Vector3 direction=node.center-center;
        if (Physics.SphereCast(center, minNodeSize * .5f,
            direction, out _, direction.magnitude)) return;
        connections.Add(node);
}
```

在八叉树中,通过双层 for 循环遍历叶节点集合寻找要构建链接的目标。如果从节点 A 的中心点向节点 B 的中心点方向形成的射线,与以节点 B 的中心点和节点大小形成的包围盒相交,并且相交的距离小于相应值,则可以判断节点 A 和 B 是相邻的,此时,节点 B 是节点 A 构建链接的目标,代码如下:

```
//第 6 章/Octree.cs

//构建叶节点间的链接
private void BuildNodeConnections()
{
    Vector3[] directionArray=new Vector3[22]
    {
            //前、后、左、右、上、下
            new Vector3(0f, 0f, 1f), new Vector3(0f, 0f, -1f),
            new Vector3(-1f, 0f, 0f), new Vector3(1f, 0f, 0f),
            new Vector3(0f, 1f, 0f), new Vector3(0f, -1f, 0f),
            //前下、前上、后下、后上
            new Vector3(0f, -1f, 1f), new Vector3(0f, 1f, 1f),
            new Vector3(0f, -1f, -1f), new Vector3(0f, 1f, -1f),
            //右上、右下、左上、左下
            new Vector3(1f, 1f, 0f), new Vector3(1f, -1f, 0f),
            new Vector3(-1f, 1f, 0f), new Vector3(-1f, -1f, 0f),
            //右前、右后、左前、左后
            new Vector3(1f, 0f, 1f), new Vector3(1f, 0f, -1f),
            new Vector3(-1f, 0f, 1f), new Vector3(-1f, 0f, -1f),
            //斜向
            new Vector3(1f, 1f, 1f), new Vector3(-1f, 1f, 1f),
            new Vector3(1f, 1f, -1f), new Vector3(-1f, 1f, -1f),
    };
    for (int i=0; i<leavesNodes.Count; i++)
    {
        OctreeNode currentNode=leavesNodes[i];
        List<OctreeNode>neighbourNodes=new List<OctreeNode>();
```

```csharp
            //遍历其他叶节点，以寻找当前节点的邻节点
            for (int j=0; j<leavesNodes.Count; j++)
            {
                if (i==j) continue;
                OctreeNode targetNode=leavesNodes[j];
                for (int k=0; k<directionArray.Length; k++)
                {
                    Vector3 direction=directionArray[k];
                    //当前节点中心向各个方向的射线
                    Ray ray=new Ray(currentNode.center, direction);
                    //射线与以目标节点中心和大小形成的包围盒是否相交
                    if (new Bounds(targetNode.center, targetNode.nodeSize
                        * Vector3.one).IntersectRay(ray, out float distance))
                    {
                        //相交时判断距离是否小于当前节点大小的一半
                        //如果小于，则将目标节点作为当前节点的邻节点
                        if (distance<currentNode.nodeSize
                            * .5f * direction.magnitude +.01f)
                            neighbourNodes.Add(targetNode);
                    }
                }
            }
            //遍历找到的邻节点，构建链接
            for (int n=0; n<neighbourNodes.Count; n++)
                currentNode.BuildConnection(neighbourNodes[n]);
        }
    }
```

通过 Gizmos 类中的 DrawLine() 方法绘制节点间的链接，代码如下：

```csharp
//第 6 章/OctreeNode.cs

if (drawCenter)
{
    Gizmos.color=Color.yellow;
    Gizmos.DrawWireSphere(center, minNodeSize * .2f);
}
if (drawConnections)
{
    if (connections !=null)
    {
```

```
            Gizmos.color=Color.cyan;
            for (int i=0; i<connections.Count; i++)
                Gizmos.DrawLine(center, connections[i].center);
    }
}
```

效果如图 6-19 所示。

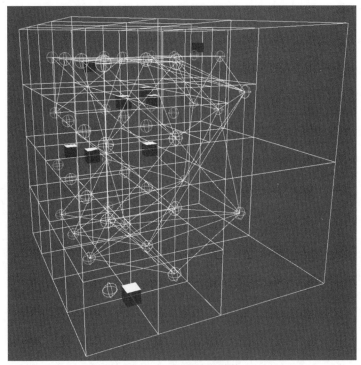

图 6-19 节点间的链接

在通过 A 星算法寻找路径时，遍历链接节点的过程便是遍历邻节点的过程，代码如下：

```
//第 6 章/Octree.cs

//<summary>
//通过 A 星算法获取路径
//</summary>
//<param name="startNode">起始节点</param>
//<param name="endNode">目标节点</param>
//<returns></returns>
public List<OctreeNode>AStar(OctreeNode startNode, OctreeNode endNode)
{
```

```csharp
if (endNode.isPassable==false) return null;
//开放集合
List<OctreeNode>openCollection=new List<OctreeNode>();
//封闭集合
List<OctreeNode>closeCollection=new List<OctreeNode>();
//将起始节点放入开放集合
openCollection.Add(startNode);
//当开放集合中的节点数量为 0 时,寻路结束
while (openCollection.Count >0)
{
    //当前节点
    OctreeNode currentNode=openCollection[0];
    //遍历查找是否有代价更小的节点
    //若代价相同,则选择与目标节点代价更小的节点
    for (int i=1; i<openCollection.Count; i++)
    {
        currentNode=(currentNode.fCost >openCollection[i].fCost
            || (currentNode.fCost==openCollection[i].fCost
                && currentNode.hCost >openCollection[i].hCost))
            ? openCollection[i] : currentNode;
    }
    //将当前节点从开放集合移除并放入封闭集合
    openCollection.Remove(currentNode);
    closeCollection.Add(currentNode);
    //如果当前节点已经是目标节点,则寻路结束
    if (currentNode==endNode) break;
    //遍历链接节点
    for (int i=0; i<currentNode.connections.Count; i++)
    {
        OctreeNode connected=currentNode.connections[i];
        //判断该链接节点是否为可通信区域,以及该链接节点是否在封闭集合中
        if (!connected.isPassable
            || closeCollection.Contains(connected))
            continue;
        //经当前节点到达该链接节点的 G 值是否小于原来的 G 值
        //或者该链接节点还没有被放入开放集合,将其放入开放集合
        float gCost=currentNode.gCost
            +CalculateCost(currentNode, connected);
        if (gCost<connected.gCost
            || !openCollection.Contains(connected))
```

```
                {
                    connected.gCost=gCost;
                    connected.hCost=CalculateCost(connected, endNode);
                    connected.pathParentNode=currentNode;
                    if (!openCollection.Contains(connected))
                        openCollection.Add(connected);
                }
            }
        }
        //倒序获取父节点
        List<OctreeNode>path=new List<OctreeNode>();
        OctreeNode currNode=endNode;
        while (currNode !=startNode)
        {
            path.Add(currNode);
            currNode=currNode.pathParentNode;
        }
        path.Add(startNode);
        path.Reverse();
        return path;
    }
```

第 7 章 游戏单位驱动

在游戏开发中,游戏单位驱动是实现游戏互动性和动态性的关键。无论是用户控制的角色、敌方战斗单位,还是游戏中的载具都需要通过精心的驱动设计展现其独特的行为和特性。

本章将详细介绍如何在 Unity 中实现各种游戏单位的驱动,包括基于 Rigidbody 组件、Character Controller 组件实现用户人物角色的驱动,基于有限状态机实现敌方战斗单位的 AI,以及载具驱动的相关内容,帮助读者掌握游戏单位驱动的技术和方法。

7.1 用户人物角色驱动

用户本身的人物角色需要根据用户的输入进行驱动,驱动可以通过 Rigidbody 组件实现,也可以通过 Character Controller 组件实现,需要依据项目的具体情况决定使用哪种组件,下面将分别介绍基于这两种组件实现人物角色驱动的方法。

7.1.1 基于刚体组件实现人物角色驱动

Rigidbody 类中包含驱使刚体移动的方法 MovePosition(),以及将刚体旋转到目标值的方法 MoveRotation(),通过这两种方法可以使刚体向目标方向移动并朝向目标方向。

在此之前,需要先获取用户的输入,根据用户的输入计算目标方向,由于相机的旋转值是不确定的,所以为了使人物角色向前移动时是向视角的前方而不是向世界空间中的前方移动,需要使用 Vector3 类中的 ProjectOnPlane()方法,获取相机的前方在地面上的投影及相机的右方在地面上的投影,这两个投影向量分别乘以用户在垂直方向和水平方向上的输入值,并将它们相加,最终归一化便是目标方向,代码如下:

```
//第 7 章/RigidbodyDriver.cs

private Camera mainCamera;
private Vector3 moveDirection;
private void Start()
```

```
{
    mainCamera=Camera.main !=null
        ?Camera.main
        :FindObjectOfType<Camera>();
}
private void Update()
{
    //获取用户输入
             ontal=Input.GetAxis("Horizontal");
                 etAxis("Vertical");

                   OnPlane(
              rd, Vector3.up) * vertical
                  ainCamera.transform.right,
                  .normalized;
```

一个向量投射到一个平面上,这个平面由方法的第 2
向,因此使用 Vector3.Up 作为法线方向表示将向量
影向量绘制出来,代码如下:

```
                      return;
                    a.transform.position;
                  3.ProjectOnPlane(
                 sform.forward, Vector3.up);
                 or3.ProjectOnPlane(
                 ansform.right, Vector3.up);
               es.DrawLine(pos, pos+fpop, 3f);
               es.DrawLine(pos, pos+rpop, 3f);
                ew GUIStyle(GUI.skin.label)
                  FontStyle.Bold
        Unity      Handles.Label(pos+fpop, "前方", style);
```

```
            UnityEditor.Handles.Label(pos +rpop, "右方", style);
        }
```

结果如图 7-1 所示。

图 7-1 投影向量

场景中的地面并非总是平坦的,假设人物角色在一个坡面上行走,便需要将移动方向投射到这个坡面上。检测地面是否为坡面,可以通过向下进行射线投射检测,根据碰撞的法线方向计算坡度,如图 7-2 所示,向量 a 与 b 的夹角是坡度,向量 n 表示法线方向,向量 u 表示正上方,法线方向与正上方之间的夹角和坡度是相等的。

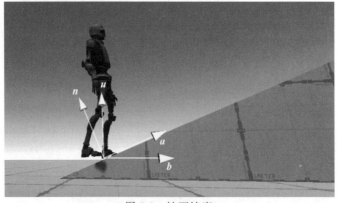

图 7-2 坡面坡度

因此计算法线方向与正上方之间的夹角便可以得知坡度,当坡度大于 0 时可以判断人物角色在坡面上,此时将移动方向投射到坡面上,代码如下:

```
//第 7 章/RigidbodyDriver.cs
```

```
[SerializeField] private LayerMask groundLayer;
private RaycastHit hitInfo;

private void Update()
{
    //...
    //坡面检测
    Ray ray=new Ray(transform.position +Vector3.up * .2f, Vector3.down);
    if (Physics.Raycast(ray, out hitInfo, .3f, groundLayer,
        QueryTriggerInteraction.Ignore))
    {
        //坡面的坡度和法线方向与世界空间上方向向量之间的角度相等
        float slopeAngle=Vector3.Angle(hitInfo.normal, Vector3.up);
        //在坡面上
        if (slopeAngle >0)
            //将移动方向投射到坡面上
            moveDirection=Vector3.ProjectOnPlane(
                moveDirection, hitInfo.normal).normalized;
    }
}
```

有了移动方向后，调用 Rigidbody 类中的 MovePosition()方法和 MoveRotation()方法设置刚体的坐标和旋转即可。当处理移动时，将移动方向乘以 Time.deltaTime 实现匀速移动，并乘以一个序列化的 float 类型的变量，用于调整移动速度。当处理旋转时，首先使用 Vector3 类中的 SignedAngle()方法计算当前方向与目标方向的角度差，然后将该角度差乘以 Time.deltaTime 及一个序列化的用于控制旋转速度的 float 类型变量，最终转换为旋转值，代码如下：

```
//第7章/RigidbodyDriver.cs

private Rigidbody rb;
[SerializeField] private float moveSpeed=2f;
[SerializeField] private float rotateSpeed=10f;

private void Start()
{
    //...
    rb=GetComponent<Rigidbody>();
}
```

```
private void Update()
{
    //...
    //移动
    rb.MovePosition(transform.position
        +Time.deltaTime * moveSpeed * moveDirection);
    //当前方向与目标方向的角度差
    float angle=Vector3.SignedAngle(
        transform.forward, moveDirection, Vector3.up);
    //旋转
    rb.MoveRotation(transform.rotation
        * Quaternion.Euler(0f, angle
            * Time.deltaTime * rotateSpeed, 0f));
}
```

7.1.2　基于角色控制器组件实现人物角色驱动

7.1.1 节中通过 Rigidbody 组件实现了人物角色的驱动，本节在此基础上将 Rigidbody 组件替换为 Character Controller 组件，使用 Character Controller 组件中的 Move()方法实现移动，该方法需要传入在移动方向上的移动增量，代码如下：

```
//第 7 章/CharacterControllerDriver.cs

using UnityEngine;

[RequireComponent(typeof(CharacterController))]
public class CharacterControllerDriver : MonoBehaviour
{
    private CharacterController cc;
    private Camera mainCamera;
    private Animator animator;
    private Vector3 moveDirection;
    [SerializeField] private float moveSpeed=2f;
    [SerializeField] private float rotateSpeed=10f;
    [SerializeField] private LayerMask groundLayer;
    private RaycastHit hitInfo;

    private void Start()
    {
        cc=GetComponent<CharacterController>();
        mainCamera=Camera.main !=null
```

```csharp
            ? Camera.main
            : FindObjectOfType<Camera>();
        animator=GetComponent<Animator>();
    }

    private void Update()
    {
        //获取用户输入
        float horizontal=Input.GetAxis("Horizontal");
        float vertical=Input.GetAxis("Vertical");
        //根据相机的朝向获取移动方向
        moveDirection=(Vector3.ProjectOnPlane(
            mainCamera.transform.forward, Vector3.up) * vertical
            +Vector3.ProjectOnPlane(mainCamera.transform.right,
            Vector3.up) * horizontal).normalized;
        //坡面检测
        Ray ray=new Ray(transform.position +Vector3.up * .2f, Vector3.down);
        if (Physics.Raycast(ray, out hitInfo, .3f, groundLayer,
            QueryTriggerInteraction.Ignore))
        {
            //坡面的坡度和法线方向与世界空间上方向向量之间的角度相等
            float slopeAngle=Vector3.Angle(hitInfo.normal, Vector3.up);
            //在坡面上
            if (slopeAngle > 0)
            {
                //将移动方向投射到坡面上
                moveDirection=Vector3.ProjectOnPlane(
                    moveDirection, hitInfo.normal).normalized;
            }
        }
        //移动
        cc.Move(Time.deltaTime * moveSpeed * moveDirection);
        //当前方向与目标方向的角度差
        float angle=Vector3.SignedAngle(
            transform.forward, moveDirection, Vector3.up);
        //旋转
        transform.rotation *=Quaternion.Euler(0f, angle
            * Time.deltaTime * rotateSpeed, 0f);
        //设置动画
        animator.SetBool("Move", moveDirection !=Vector3.zero);
    }
}
```

Character Controller 组件与 Rigidbody 组件不同的是,后者可以模拟真实的物理行为,具有重力效果,而使用前者驱动人物角色需要自行实现重力效果,否则当角色从高处走出时将悬浮于空中,如图 7-3 所示。

图 7-3 人物角色悬浮于空中

可以通过向下进行球体投射检测,根据是否有碰撞判断是否处于地面,球体投射检测的半径参考角色控制器的半径,代码如下:

```
//第 7 章/CharacterControllerDriver.cs

private bool isGrounded;
[SerializeField] private float groundCheckOffset=.03f;
private Vector3 groundNormal;

//地面检测
private void GroundCheck()
{
    float radius=cc.radius;
    Vector3 origin=transform.position
        +(radius +groundCheckOffset) * Vector3.up;
    if (Physics.SphereCast(origin, radius, Vector3.down,
        out RaycastHit hitInfo, groundCheckOffset * 2f,
        groundLayer, QueryTriggerInteraction.Ignore))
    {
        isGrounded=true;
        groundNormal=hitInfo.normal;
    }
    else isGrounded=false;
}
```

定义重力值，当人物角色处于空中时，根据重力值不断地加大垂直方向上下落的向量，最终将该向量添加给移动增量，代码如下：

```
//第7章/CharacterControllerDriver.cs

[SerializeField] private float gravity=-9.81f;
private float verticalVelocity;

private void Update()
{
    //...
    GroundCheck();
    GravityApply();
    //移动
    cc.Move(Time.deltaTime * moveSpeed * moveDirection
        +Time.deltaTime * verticalVelocity * Vector3.up);
    //...
}
//重力应用
private void GravityApply()
{
    if (isGrounded)
        verticalVelocity=Mathf.Lerp(
            gravity, -2f, Vector3.Dot(Vector3.up, groundNormal));
    else
        verticalVelocity+=gravity * Time.deltaTime;
}
```

假设人物角色从高空掉落到一个斜面上，此时地面检测的球体投射检测结果为true，如果这个斜面的坡度大于角色控制器设定的坡度限制值，则此时判定人物角色处于地面显然是不合适的，应该让其沿坡面继续掉落，因此还需要根据碰撞的法线方向求得坡度，从而进一步地进行判断，代码如下：

```
//第7章/CharacterControllerDriver.cs

//地面检测
private void GroundCheck()
{
    float radius=cc.radius;
    Vector3 origin=transform.position
        +(radius +groundCheckOffset) * Vector3.up;
```

```
        if (Physics.SphereCast(origin, radius, Vector3.down,
            out RaycastHit hitInfo, groundCheckOffset * 2f,
            groundLayer, QueryTriggerInteraction.Ignore))
        {
            groundNormal=hitInfo.normal;
            float angle=Vector3.Angle(hitInfo.normal, Vector3.up);
            isGrounded=angle<=cc.slopeLimit;
        }
        else isGrounded=false;
}
```

计算让人物沿坡面继续掉落的向量需要用到向量的叉乘运算, 2.2.4 节中介绍了向量 *a* 与向量 *b* 的叉乘运算结果是另一个向量, 该向量垂直于向量 *a* 和向量 *b* 构成的平面, 因此, 通过碰撞的法线方向和正上方向量进行叉乘运算, 再将叉乘运算结果与碰撞的法线方向进行叉乘运算, 得到的便是目标向量, 最终将其应用到移动增量, 代码如下:

```
//第 7 章/CharacterControllerDriver.cs

private Vector3 slopeVelocity;

private void Update()
{
    //...
    GroundCheck();
    GravityApply();
    //移动
    cc.Move(Time.deltaTime * moveSpeed * moveDirection
        +Time.deltaTime * verticalVelocity * Vector3.up
        +Time.deltaTime * slopeVelocity);
    //...
}

//地面检测
private void GroundCheck()
{
    Vector3 slopeVelocity=Vector3.zero;
    float radius=cc.radius;
    Vector3 origin=transform.position
        +(radius +groundCheckOffset) * Vector3.up;
    if (Physics.SphereCast(origin, radius, Vector3.down,
        out RaycastHit hitInfo, groundCheckOffset * 2f,
```

```csharp
            groundLayer, QueryTriggerInteraction.Ignore))
{
    groundNormal=hitInfo.normal;
    float angle=Vector3.Angle(hitInfo.normal, Vector3.up);
    isGrounded=angle<=cc.slopeLimit;
    if (!isGrounded)
    {
        Vector3 slopeRight=Vector3.Cross(
            hitInfo.normal, Vector3.up).normalized;
        slopeVelocity=Vector3.Cross(
            hitInfo.normal, slopeRight).normalized;
        Debug.DrawLine(hitInfo.point, hitInfo.point +hitInfo.normal);
        Debug.DrawLine(hitInfo.point, hitInfo.point +Vector3.up);
        Debug.DrawLine(hitInfo.point, hitInfo.point +slopeRight);
        Debug.DrawLine(hitInfo.point, hitInfo.point +slopeVelocity);
    }
}
else isGrounded=false;

if (slopeVelocity.magnitude >.001f)
    this.slopeVelocity+=Time.deltaTime
        * Mathf.Abs(gravity) * slopeVelocity;
else
    this.slopeVelocity=Vector3.Lerp(this.slopeVelocity,
        Vector3.zero, Time.deltaTime * 20f);
}
```

Debug.DrawLine()方法的绘制结果如图7-4所示。

图7-4 绘制结果

7.2 人物角色行为

7.1节介绍了人物角色驱动的实现过程，本节在此基础上，介绍如何为人物角色实现各种行为，包括跳跃、滑行、翻越和掩体行为。这些行为不仅可以为人物角色增添生动性和真实感，还能极大地丰富游戏的玩法和策略性。

7.2.1 跳跃

跳跃过程分为3个阶段，分别是起跳、下降和落地，状态结构如图7-5所示。跳跃过程需要如此划分是由于过渡到下降状态的源状态并非只有起跳，假设人物角色从一个高处走出，受重力的影响，人物角色将从空中降落到地面，因此移动和起跳状态均可过渡到下降状态。

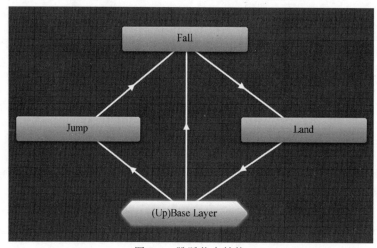

图 7-5　跳跃状态结构

7.1.2节中通过Character Controller组件实现了人物角色的驱动，其中包括地面的检测和重力的应用，本节在此基础上增加跳跃功能。

假设通过键盘空格按键触发跳跃，那么在isGrounded变量为true时，判断空格按键是否被按下。为了避免在跳跃过程中重复触发跳跃，可以增加冷却计时变量。

当触发跳跃时，根据跳跃动力值设置垂直方向上的移动量，并播放起跳动画。下降通过计时实现，当在空中经过指定的时长后，进入下降状态，代码如下：

```
//第7章/AvatarController.cs

//跳跃动力值,值越大跳跃高度越高
[SerializeField] private float jumpPower=1.2f;
//跳跃冷却时长
[SerializeField] private float jumpCD=1f;
```

```csharp
//在空中经过该时长进入下降状态
[SerializeField] private float fallTime=.2f;
//跳跃和下降的计时器
private float jumpTimer, fallTimer;

//重力应用
private void GravityApply()
{
    if (isGrounded)
    {
        animator.SetBool(AnimParam.Land, true);
        fallTimer=fallTime;

        verticalVelocity=Mathf.Lerp(
            gravity, -2f, Vector3.Dot(Vector3.up, groundNormal));

        animator.SetBool(AnimParam.Jump, false);
        animator.SetBool(AnimParam.Fall, false); //下降动画
        //触发跳跃
        if (Input.GetKeyDown(KeyCode.Space) && jumpTimer<=0f)
        {
            jumpTimer=jumpCD;    //重置跳跃计时
            verticalVelocity=Mathf.Sqrt(jumpPower * -2f * gravity);
            animator.SetBool(AnimParam.Jump, true); //起跳动画
        }
        if (jumpTimer >= 0f)
            jumpTimer -=Time.deltaTime;
    }
    else
    {
        //下降
        if (fallTimer >0f)
            fallTimer -=Time.deltaTime;
        else
            animator.SetBool(AnimParam.Fall, true);

        verticalVelocity+=gravity * Time.deltaTime;
        //落地动画
        animator.SetBool(AnimParam.Land, false);
    }
}
```

跳跃的 3 个阶段如图 7-6 所示。

（a）起跳

（b）下降

（c）落地

图 7-6　跳跃的 3 个阶段

7.2.2　滑行

本节实现人物角色的滑行功能，首先搭建一个简易的场景，如图 7-7 所示，使用 3 个大小不一的立方体作为触发对象，触发对象的下方是滑行区域。新增 1 个标签，将其命名为 Slide，并为这些立方体设置该标签。

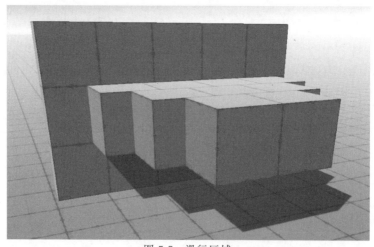

图 7-7　滑行区域

通常情况下，人物角色的轴心点在其底部，即两只脚的中点，如图 7-8 所示。在人物角色的坐标加上方一定单位处向人物角色的前方进行射线投射检测，当检测到标签为 Slide 的物体时，如果满足滑行条件，则触发滑行动作。

为滑行及其他特殊的人物动作编写一个基类，基类中提供用于检测当前是否满足动作触发条件、动作执行等抽象方法或虚方法，以及相关的变量，例如射线投射检测的高度、距离、触发对象的标签、可滑行的最大距离等，代码如下：

图 7-8 人物角色的轴心点

```
//第7章/AvatarAction.cs

using UnityEngine;

public abstract class AvatarAction
{
    protected AvatarController controller;
    protected float raycastHeight;
    protected float raycastDistance;
    protected float actionMaxDistance;
    protected string targetTag;
    protected Vector3 startPosition, targetPosition;
    protected Quaternion startRotation, targetRotation;
    protected float timer;
    protected float actionDuration;

    public AvatarAction(AvatarController controller,
        AvatarActionSettings settings)
    {
        this.controller=controller;
        raycastHeight=settings.raycastHeight;
        raycastDistance=settings.raycastDistance;
        targetTag=settings.targetTag;
        actionDuration=settings.actionDuration;
        actionMaxDistance=settings.actionMaxDistance;
    }
```

```
    public abstract bool ActionDoableCheck();
    public abstract bool ActionExecute();
    public virtual void OnAnimatorIK(int layerIndex) { }
    public virtual void OnDrawGizmos() { }
}
```

在人物角色动作类的构造函数中，需要提供用于该动作的配置参数，通过继承 ScriptableObject 类实现配置的编辑和存储，代码如下：

```
//第 7 章/AvatarActionSettings.cs

using UnityEngine;

[CreateAssetMenu]
public class AvatarActionSettings : ScriptableObject
{
    public float raycastHeight;
    public float raycastDistance;
    public string targetTag;
    public float actionDuration;
    public float actionMaxDistance;
}
```

假设在人物角色上方 1.5m 位置处向前方进行射线投射检测，检测的长度为 1.5m，滑行动作的时长为 1s，滑行的最大距离为 4m，配置如图 7-9 所示。

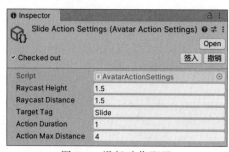

图 7-9　滑行动作配置

通常情况下，人物角色只能沿着物体的一个轴向滑行，此处以 z 轴为例，当检测到标签为 Slide 的碰撞体后，在碰撞信息中获取法线向量，将法线向量与碰撞体的前方向量做点乘运算，通过点乘结果可以得知它们之间的夹角大小，如果夹角过大，则当前不满足滑行条件。

同理，还可以通过法线向量与人物角色的前方向量做点乘运算，进而判断人物角色是否面向碰撞体，代码如下：

```csharp
//第7章/SlideAction.cs

public override bool ActionDoableCheck()
{
    if (!controller.IsGrounded) return false;
    //在指定高度处向前进行射线投射检测
    Vector3 origin=controller.transform.position
        +raycastHeight * Vector3.up;
    if (Physics.Raycast(origin, controller.transform.forward,
        out RaycastHit hitInfo, raycastDistance))
    {
        if (!hitInfo.collider.CompareTag(targetTag)) return false;
        //法线反方向和角色前方的向量叉积
        float dot=Vector3.Dot(-hitInfo.normal,
            controller.transform.forward);
        float angle=Mathf.Acos(dot) * Mathf.Rad2Deg;
        if (angle >45f) return false; //角度过大
        //法线方向和碰撞物体前方的向量叉积
        dot=Vector3.Dot(hitInfo.normal,
            hitInfo.transform.forward);
        if (Mathf.Abs(dot)<.8f) return false; //与碰撞物体 Z 轴方向角度相差过大
        Debug.DrawLine(hitInfo.point,
            hitInfo.point +hitInfo.normal, Color.blue);
        return true;
    }
    return false;
}
```

除此之外，还需要判断碰撞体在 z 轴方向上的大小是否等于人物角色可滑行的最大距离，如果场景中用于触发滑行动作的物体是由大小为 1 的 Cube 拉伸而成的，则可以使用 Transform 中 lossyScale 的 z 值判断，但是仅适用于 Cube，通常情况下，它不能用于表示真实大小。为了得到碰撞体在 z 轴方向上的大小，需要用到 Bounds 类中的 Encapsulate()方法，它用于扩充包围盒，以包含其他的包围盒或点，代码如下：

```csharp
//第7章/SlideAction.cs

//碰撞体在 z 轴上的大小
Bounds bounds=new Bounds();
var mf=hitInfo.collider.GetComponent<MeshFilter>();
bounds.Encapsulate(mf.sharedMesh.bounds);
float distance=bounds.size.z;
if (distance >actionMaxDistance) return false; //距离过大
```

满足以上条件后,沿着法线反方向在碰撞体 z 轴大小加一定偏移量处向下进行射线投射检测,只检测表示地面的层级物体,检测到的碰撞点即是滑行的目标点,代码如下:

```
//第7章/SlideAction.cs

//沿法线反方向在碰撞体 z 轴大小加一定偏移量处向下进行射线投射检测
origin+=hitInfo.point -origin +-hitInfo.normal * (distance +.8f);
if (Physics.Raycast(origin, Vector3.down, out RaycastHit hitInfo2,
    raycastHeight +.01f, 1 <<LayerMask.NameToLayer("Ground")))
{
    timer=0f;
    startPosition=controller.transform.position;
    startRotation=controller.transform.rotation;
    targetPosition=hitInfo2.point;
    targetRotation=Quaternion.LookRotation(
        targetPosition -startPosition);
    controller.EnableCollider(false);
    controller.animator.SetTrigger("Slide");
    return true;
}
```

在动作执行的过程中,通过一个 float 类型的变量进行计时,当其大于或等于动作的时长时,表示动作执行完成,方法的返回值为 false。未执行完成前,从起始位置和朝向向目标位置和朝向做插值运算,并更新人物角色的位置和朝向,代码如下:

```
//第7章/SlideAction.cs

public override bool ActionExecute()
{
    timer+=Time.deltaTime;
    if (timer >actionDuration)
        timer=actionDuration;
    float t=timer / actionDuration;
    controller.transform.SetPositionAndRotation(
        Vector3.Lerp(startPosition, targetPosition, t),
        Quaternion.Lerp(startRotation, targetRotation, t));
    return timer<actionDuration;
}
```

效果如图 7-10 所示。

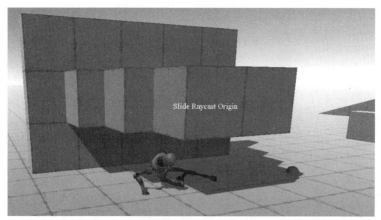

图 7-10　人物角色滑行动作

7.2.3　翻越

翻越动作的实现过程与滑行动作基本一致，不同的是，由于翻越动作在执行过程中左手手部需要支持在障碍物上方，如图 7-11 所示，因此需要加入 IK 的应用，使左手贴合在障碍物上方。

图 7-11　人物角色翻越动作

左手的目标 IK 位置在检测到当前满足动作的执行条件后进行计算，从碰撞信息中获取碰撞位置和法线方向，将碰撞位置沿法线反方向移动碰撞体在 z 轴方向上大小的一半，得到在 z 轴方向上的中点。包围盒中心点的 y 值加 y 轴方向上大小的一半，便是目标 IK 位置的高度，为了避免穿模现象，还应该考虑手掌的厚度。正上方和碰撞的法线方向两个向量做叉乘运算，可以得到以人物角色的前方为观察方向的碰撞体的左侧方向，再沿着该方向移动一定距离便可以得到最终的目标 IK 位置，移动的距离根据手臂的长度做适当调整，代码如下：

```csharp
//第 7 章/VaultAction.cs

public override bool ActionDoableCheck()
{
    if (!controller.IsGrounded) return false;
    Vector3 origin=controller.transform.position
        +raycastHeight * Vector3.up;
    if (Physics.Raycast(origin, controller.transform.forward,
        out RaycastHit hitInfo, raycastDistance))
    {
        if (!hitInfo.collider.CompareTag(targetTag)) return false;
        if (hitInfo.distance<=.8f) return false;
        float dot=Vector3.Dot(-hitInfo.normal,
            controller.transform.forward);
        float angle=Mathf.Acos(dot) * Mathf.Rad2Deg;
        if (angle >25f) return false;
        dot=Vector3.Dot(hitInfo.normal,
            hitInfo.transform.forward);
        if (Mathf.Abs(dot)<.8f) return false;
        Debug.DrawLine(hitInfo.point,
            hitInfo.point +hitInfo.normal, Color.blue);
        Bounds bounds=new Bounds();
        var mf=hitInfo.collider.GetComponent<MeshFilter>();
        bounds.Encapsulate(mf.sharedMesh.bounds);
        float distance=bounds.size.z;
        if (distance >actionMaxDistance) return false;
        origin+=hitInfo.point -origin +-hitInfo.normal * (distance +1f);
        if (Physics.Raycast(origin, Vector3.down, out RaycastHit hitInfo2,
            raycastHeight +.01f, 1 <<LayerMask.NameToLayer("Ground")))
        {
            timer=0f;
            startPosition=controller.transform.position;
            startRotation=controller.transform.rotation;
            targetPosition=hitInfo2.point;
            targetRotation=Quaternion.LookRotation(
                targetPosition -startPosition);
            controller.EnableCollider(false);
            controller.animator.SetTrigger("Vault");
            //计算左手 IK 目标位置和旋转
            leftHandIKPosition=hitInfo.point +-hitInfo.normal
                * bounds.size.z * .5f; //Z 轴上取中点
```

```
        //Y轴上取最高点,并加一定偏移量,手掌越厚偏移值应越大
        leftHandIKPosition.y=bounds.center.y
            +bounds.size.y * .5f +.08f;
        //根据叉乘求得左侧
        Vector3 left=Vector3.Cross(Vector3.up, hitInfo.normal);
        leftHandIKPosition+=left * .6f; //向左侧偏移一定单位
        leftHandIKRotation=Quaternion.LookRotation(
            -hitInfo.normal, Vector3.up);
        return true;
    }
}
targetPosition=Vector3.zero;
return false;
}
```

求得左手的目标 IK 位置和旋转后,在 OnAnimatorIK()方法中进行应用,需要注意的是,IK 的权重值需要依据动画播放进度进行设置。

新增一条动画曲线,作为左手的 IK 权重值曲线,在到达左手着力点之前,权重值逐渐过渡到 1,如图 7-12(a)所示,之后再逐渐恢复为 0,如图 7-12(b)所示。

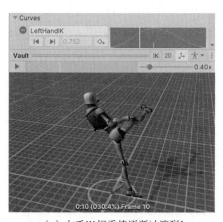
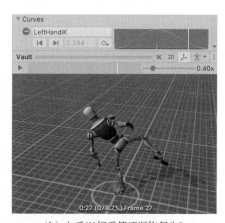

(a)左手IK权重值逐渐过渡到1　　　(b)左手IK权重值逐渐恢复为0

图 7-12　左手 IK 权重值曲线

在 Animator 中新增 1 个与左手 IK 权重值曲线同名的参数值,在 OnAnimatorIK()方法中读取该参数值,并将其作为左手 IK 权重值,代码如下:

```
//第 7 章/VaultAction.cs

public override void OnAnimatorIK(int layerIndex)
{
```

```csharp
        float leftHandIK=controller.animator
            .GetFloat("LeftHandIK");
        controller.animator.SetIKPositionWeight(
            AvatarIKGoal.LeftHand, leftHandIK);
        controller.animator.SetIKPosition(
            AvatarIKGoal.LeftHand, leftHandIKPosition);
        controller.animator.SetIKRotationWeight(
            AvatarIKGoal.LeftHand, leftHandIK);
        controller.animator.SetIKRotation(
            AvatarIKGoal.LeftHand, leftHandIKRotation);
}
```

使用Gizmos标注射线投射检测的起点、翻越的目标位置、左手的目标IK位置等信息，便于调试，代码如下：

```csharp
//第7章/VaultAction.cs

public override void OnDrawGizmos()
{
    base.OnDrawGizmos();
    Color cacheColor=Gizmos.color;
    Gizmos.color=Color.magenta;
    //射线投射检测的起点
    Vector3 origin=controller.transform.position
        +raycastHeight * Vector3.up;
    Gizmos.DrawWireSphere(origin, .08f);

#if UNITY_EDITOR
    GUIStyle style=new GUIStyle(GUI.skin.label) { fontSize=15 };
    UnityEditor.Handles.Label(origin +Vector3.right * .1f,
        "Vault Raycast Origin", style);
#endif
    if (targetPosition !=Vector3.zero)
    {
        Gizmos.DrawSphere(targetPosition, .13f);
#if UNITY_EDITOR
        UnityEditor.Handles.Label(targetPosition +Vector3.right * .1f,
            "Vault Target Position", style);

        //在左手目标IK位置绘制箭头
        UnityEditor.Handles.ArrowHandleCap(0, leftHandIKPosition,
            Quaternion.LookRotation(Vector3.up), .5f, EventType.Repaint);
```

```
        UnityEditor.Handles.Label(leftHandIKPosition,
            "Left Hand IK Position", style);
#endif
    }
    Gizmos.color=cacheColor;
}
```

效果如图 7-13 所示。

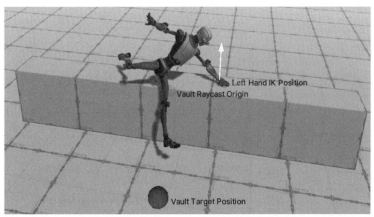

图 7-13 使用 Gizmos 调试人物角色翻越动作

7.2.4 掩体行为

掩体行为相较于滑行和翻越行为要复杂一些，因为掩体行为不止有一个动作，包含从普通站立状态切换至掩体状态的动作、掩体静止动作及掩体潜行等动作，如图 7-14 所示。

（a）进入掩体　　　　　　　（b）掩体静止　　　　　　　（c）掩体潜行

图 7-14 人物角色掩体行为

在掩体动作类中声明一个枚举,用于表示掩体行为中的不同动作状态,当重写检测动作是否为可执行的方法时,检测当前是否可以切换至掩体状态。

由于掩体行为的触发对象的高度是不定的,并且人物角色会依据掩体的高度调整身体的高度,因此在进行物理检测时使用射线投射检测不再合适,改为使用盒体投射检测,为了实现掩体动作相关参数的配置和存储,新建一个配置类,代码如下:

```
//第 7 章/CoverActionSettings.cs

using UnityEngine;

[CreateAssetMenu]
public class CoverActionSettings : AvatarActionSettings
{
    public Vector3 boxCastSize=Vector3.one * .4f;
    public float sneakSpeed=1f;
    public float directionLerpSpeed=3f;
    public float headRadius=.12f;
    public int headDownCastCountLimit=3;
    public float bodyPositionLerpSpeed=0.05f;
    public float footPositionLerpSpeed=0.15f;
}
```

当处于掩体状态时,获取水平方向上的输入;当输入值小于 0 时,向左侧潜行;当输入值大于 0 时,向右侧潜行;当输入值为 0 时,人物角色保持静止状态。

为了实现从静止到潜行的平滑过渡,使用混合树搭建动画状态机,如图 7-15 所示,混合树分为两个,代表不同的方向,尽管它们分为两个,但是可以使用相同的动画剪辑,通过启用镜像动画将动画调整为相反的方向,如图 7-16 所示。

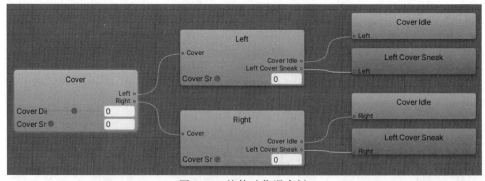

图 7-15 掩体动作混合树

在潜行的过程中,由于人物角色背后的掩体并不一定始终是平坦的,如图 7-17 所示,因此为了让人物角色始终背靠掩体,需要向背后进行射线投射检测,移动时沿着碰撞的法线反

图 7-16　镜像动画

图 7-17　圆柱形掩体

方向进行移动,代码如下:

```csharp
//第 7 章/CoverAction.cs

public override bool ActionExecute()
{
    switch (state)
    {
        case State.StandToCover:
            timer+=Time.deltaTime;
            if (timer >actionDuration)
                timer=actionDuration;
            float t=timer / actionDuration;
            controller.transform.SetPositionAndRotation(
                Vector3.Lerp(startPosition, targetPosition, t),
                Quaternion.Lerp(startRotation, targetRotation, t));
            if (t==1)
                state=State.Covering;
            break;
        case State.Covering:
            controller.EnableCollider(true);
```

```csharp
/***********************************************************
 * 根据水平方向的输入进行左右移动
 * 在移动的过程中向身体后方进行射线投射检测
 * 目的是获取碰撞的法线方向,使角色移动时始终背靠掩体对象
 ***********************************************************/
float horizontal=Input.GetAxis("Horizontal");
if (horizontal !=0f)
{
    hInputSign=horizontal >0f ? 1f : -1f;
    Physics.Raycast(controller.transform.position +Vector3.up
        * raycastHeight, controller.transform.forward,
        out RaycastHit hit, 0.5f);
    Debug.DrawLine(hit.point, hit.point +hit.normal,
        Color.magenta);
    cc.Move(-hit.normal * sneakSpeed * Time.deltaTime);
    controller.transform.rotation=Quaternion.Lerp(
        controller.transform.rotation,
        Quaternion.LookRotation(-hit.normal),
        Time.deltaTime * 5f);
}

//设置动画参数值
coverDirection=Mathf.Lerp(coverDirection,
    hInputSign, Time.deltaTime * dirLerpSpeed);
controller.animator.SetFloat("Cover Direction",
    coverDirection);
controller.animator.SetFloat("Cover Sneak",
    Mathf.Abs(horizontal));

//当垂直方向上的输入小于 0 时退出掩体状态
float vertical=Input.GetAxis("Vertical");
if (vertical< 0f)
{
    timer=0f;
    startPosition=controller.transform.position;
    startRotation=controller.transform.rotation;
    Physics.Raycast(controller.transform.position +Vector3.up
        * raycastHeight, controller.transform.forward,
        out RaycastHit hit, 0.5f);
    targetPosition=controller.transform.position
        +hit.normal * (raycastDistance +.1f);
```

```
                    targetRotation=Quaternion.LookRotation(-hit.normal);
                    state=State.CoverToStand;
                    controller.animator.SetBool("Cover", false);
                    controller.EnableCollider(false);
                }
                break;
            case State.CoverToStand:
                timer+=Time.deltaTime;
                if (timer >actionDuration)
                    timer=actionDuration;
                t=timer / actionDuration;
                controller.transform.SetPositionAndRotation(
                    Vector3.Lerp(startPosition, targetPosition, t),
                    Quaternion.Lerp(startRotation, targetRotation, t));
                return timer<actionDuration;
        }
        return true;
    }
}
```

在潜行的过程中,当掩体的高度发生变化时,人物角色应该降低身体高度,如图7-18所示,为了避免在降低身体高度的过程中脚部发生穿模现象,需要启用脚部的IK。

(a) 未降低身体高度　　　　　　　　　(b) 降低身体高度

图 7-18　根据掩体调整人物角色高度

如何判断掩体的高度发生了变化? 可以通过从头部的位置向后方进行球体投射检测,如果未检测到任何碰撞体,则说明头部的高度已经高于掩体的高度。

当向左侧潜行时,向头的左后方进行球体投射检测,相应地,当向右侧潜行时,向头的右后方进行球体投射检测,代码如下:

//第7章/CoverAction.cs

```
/****************************************************************
 * 向头的左后方或右后方进行球体投射检测
 * 目的是当检测不到碰撞体时向下调整身体高度并使用脚部 IK
 * 如果在头部初始高度检测不到,则每次下降一个单位再检测
 ****************************************************************/
bool headCastResult=false;
for (int i=0; i<headDownCastCountLimit; i++)
{
    headSphereCastOrigin=controller.transform.position
        +Vector3.up * (headOriginPosY - i * headRadius * 2)
        +controller.transform.right * hInputSign * headRadius * 2;
    headCastResult=Physics.SphereCast(headSphereCastOrigin,
        headRadius, controller.transform.forward, out headHitInfo);
    if (headCastResult) break;
}
if (headCastResult)
{
    targetBodyPositionY=originBodyPositionY
        -(headOriginPosY -headHitInfo.point.y) -headRadius;
    cc.Move(horizontal * sneakSpeed * Time.deltaTime
        * controller.transform.right);
}
```

从脚部的上方一定单位处向下进行射线投射检测,通过碰撞信息获取脚部的目标 IK 位置和旋转,代码如下:

```
//第 7 章/CoverAction.cs

/****************************************************************
 * 从脚部的上方一定单位处向下进行射线投射检测
 * 目的是获取脚部的目标 IK 位置和旋转
 ****************************************************************/
Vector3 leftFootPosition=controller.animator
    .GetBoneTransform(HumanBodyBones.LeftFoot).position;
leftFootPosition.y=controller.transform.position.y +raycastHeight;
if (Physics.Raycast(leftFootPosition, Vector3.down,
    out RaycastHit footHitInfo, raycastHeight +.01f))
{
    leftFootIkPosition=footHitInfo.point;
    leftFootIkRotation=Quaternion.FromToRotation(
        controller.transform.up, footHitInfo.normal);
```

```
}
else leftFootIkPosition=Vector3.zero;
Vector3 rightFootPosition=controller.animator
    .GetBoneTransform(HumanBodyBones.RightFoot).position;
rightFootPosition.y=controller.transform.position.y +raycastHeight;
if (Physics.Raycast(rightFootPosition, Vector3.down,
    out footHitInfo, raycastHeight +.01f))
{
    rightFootIkPosition=footHitInfo.point;
    rightFootIkRotation=Quaternion.FromToRotation(
        controller.transform.up, footHitInfo.normal);
}
else rightFootIkPosition=Vector3.zero;
```

计算出身体的目标高度和脚部的目标 IK 位置和旋转后,在 OnAnimatorIK() 方法中进行应用,代码如下：

```
//第7章/CoverAction.cs

public override void OnAnimatorIK(int layerIndex)
{
    if (state !=State.Covering) return;
    if (originBodyPositionY==0)
    {
        originBodyPositionY=controller.animator.bodyPosition.y;
        targetBodyPositionY=controller.animator.bodyPosition.y;
        lastBodyPositionY=originBodyPositionY;
        return;
    }
    Vector3 bodyPosition=controller.animator.bodyPosition;
    bodyPosition.y=Mathf.Lerp(lastBodyPositionY,
        targetBodyPositionY, bodyPositionLerpSpeed);
    controller.animator.bodyPosition=bodyPosition;
    lastBodyPositionY=controller.animator.bodyPosition.y;
    //左脚 IK
    controller.animator.SetIKPositionWeight(AvatarIKGoal.LeftFoot, 1f);
    controller.animator.SetIKRotationWeight(AvatarIKGoal.LeftFoot, 1f);
    Vector3 targetIkPosition=controller.animator
        .GetIKPosition(AvatarIKGoal.LeftFoot);
    if (leftFootIkPosition !=Vector3.zero)
    {
```

```csharp
            targetIkPosition=controller.transform
                .InverseTransformPoint(targetIkPosition);
            Vector3 world2Local=controller.transform
                .InverseTransformPoint(leftFootIkPosition);
            float y=Mathf.Lerp(lastLeftFootPositionY,
                world2Local.y, footPositionLerpSpeed);
            targetIkPosition.y+=y;
            lastLeftFootPositionY=y;
            targetIkPosition=controller.transform
                .TransformPoint(targetIkPosition);
            Quaternion currRotation=controller.animator
                .GetIKRotation(AvatarIKGoal.LeftFoot);
            Quaternion nextRotation=leftFootIkRotation * currRotation;
            controller.animator.SetIKRotation(
                AvatarIKGoal.LeftFoot, nextRotation);
        }
        controller.animator.SetIKPosition(
            AvatarIKGoal.LeftFoot, targetIkPosition);
        //右脚IK
        controller.animator.SetIKPositionWeight(AvatarIKGoal.RightFoot, 1f);
        controller.animator.SetIKRotationWeight(AvatarIKGoal.RightFoot, 1f);
        targetIkPosition=controller.animator
            .GetIKPosition(AvatarIKGoal.RightFoot);
        if (rightFootIkPosition !=Vector3.zero)
        {
            targetIkPosition=controller.transform
                .InverseTransformPoint(targetIkPosition);
            Vector3 world2Local=controller.transform
                .InverseTransformPoint(rightFootIkPosition);
            float y=Mathf.Lerp(lastRightFootPositionY,
                world2Local.y, footPositionLerpSpeed);
            targetIkPosition.y+=y;
            lastRightFootPositionY=y;
            targetIkPosition=controller.transform
                .TransformPoint(targetIkPosition);
            Quaternion currRotation=controller.animator
                .GetIKRotation(AvatarIKGoal.RightFoot);
            Quaternion nextRotation=rightFootIkRotation * currRotation;
            controller.animator.SetIKRotation(
                AvatarIKGoal.RightFoot, nextRotation);
        }
```

```
        controller.animator.SetIKPosition(
            AvatarIKGoal.RightFoot, targetIkPosition);
    }
```

7.3 敌方战斗单位驱动

敌方战斗单位的移动过程通常是自动寻路的过程,而触发其寻路行为的通常是敌方战斗单位的 AI。例如,一个敌方战斗单位有 3 种状态,分别为巡逻状态、追击状态、攻击状态。巡逻状态为默认状态,在该状态下,敌方战斗单位在指定的巡逻点之间巡逻走动,同时检测 Player 的位置,当 Player 进入其警戒范围时,进入追击状态,朝 Player 移动,当与 Player 的距离小于其攻击范围时,进入攻击状态,当 Player 离开攻击范围但仍在警戒范围内时,敌方战斗单位将切换至追击状态,继续追击,当 Player 离开警戒范围时,敌方战斗单位将回到巡逻状态。本节将介绍有限状态机的相关内容,并介绍如何基于有限状态机实现这个敌方战斗单位的 AI。

7.3.1 有限状态机

有限状态机(Finite State Machine,FSM)是一种用来为对象行为进行建模的工具。它表示有限种状态及在这些状态之间进行切换等行为的模型。

有限状态机包含状态和状态机两块核心内容。状态中包含各种状态事件,例如进入状态时执行的进入事件,退出状态时执行的退出事件等。在本节中,为状态类声明 5 个基本状态事件,分别是状态初始化事件、状态进入事件、状态停留事件、状态退出事件和状态终止事件。除此之外,还包含一个用于设置状态切换条件的方法,当条件满足时,状态机将切换至目标状态,代码如下:

```
//第7章/State.cs

using System;

public class State : IState
{
    public string Name { get; set; }

    public StateMachine machine;
    public Action onInitialization;
    public Action onEnter;
    public Action onStay;
    public Action onExit;
```

```csharp
    public Action onTermination;

    public virtual void OnInitialization()
    {
        onInitialization?.Invoke();
    }
    public virtual void OnEnter()
    {
        onEnter?.Invoke();
    }
    public virtual void OnStay()
    {
        onStay?.Invoke();
    }
    public virtual void OnExit()
    {
        onExit?.Invoke();
    }
    public virtual void OnTermination()
    {
        onTermination?.Invoke();
    }
    //<summary>
    //设置状态切换条件
    //</summary>
    //<param name="predicate">切换条件</param>
    //<param name="targetStateName">目标状态名称</param>
    public void SwitchWhen(Func<bool>predicate, string targetStateName)
    {
        machine.SwitchWhen(predicate, Name, targetStateName);
    }
}
```

状态机用于管理所有的状态,管理行为包括添加状态、移除状态、获取状态及切换状态等。当添加状态时,将状态添加到状态存储列表,并执行其初始化事件。当移除状态时,将其从状态存储列表移除,并执行其终止事件。当切换状态时,首先会执行当前状态的退出事件,然后将当前状态更新为目标状态,更新后,执行当前状态的进入事件。

当状态机刷新过程中,除了执行当前状态的停留事件外,还将检测所有的状态切换条件。当条件满足时,状态机自动切换至目标状态,状态切换条件通过SwitchWhen()方法设置,代码如下:

```csharp
//第7章/StateMachine.cs
using System;
using System.Collections.Generic;

public class StateMachine
{
    protected readonly List<IState> states=new List<IState>();
    protected List<StateSwitchCondition> conditions
        =new List<StateSwitchCondition>();
    public IState CurrentState { get; protected set; }

    public bool Add(IState state)
    {
        //判断是否已经存在
        if (!states.Contains(state))
        {
            //判断是否存在同名状态
            if (states.Find(m =>m.Name==state.Name)==null)
            {
                //存储到列表
                states.Add(state);
                //执行状态初始化事件
                state.OnInitialization();
                return true;
            }
        }
        return false;
    }
    public bool Add<T>(string stateName=null) where T : IState, new()
    {
        Type type=typeof(T);
        T t=(T)Activator.CreateInstance(type);
        t.Name=string.IsNullOrEmpty(stateName) ? type.Name : stateName;
        return Add(t);
    }
    public bool Remove(IState state)
    {
        //判断是否存在
        if (states.Contains(state))
        {
```

```csharp
            //如果要移除的状态为当前状态,则首先执行当前状态退出事件
            if (CurrentState==state)
            {
                CurrentState.OnExit();
                CurrentState=null;
            }
            //执行状态终止事件
            state.OnTermination();
            return states.Remove(state);
        }
        return false;
    }
    public bool Remove(string stateName)
    {
        var targetIndex=states.FindIndex(m =>m.Name==stateName);
        if (targetIndex!=-1)
        {
            var targetState=states[targetIndex];
            if (CurrentState==targetState)
            {
                CurrentState.OnExit();
                CurrentState=null;
            }
            targetState.OnTermination();
            return states.Remove(targetState);
        }
        return false;
    }
    public bool Remove<T>() where T : IState
    {
        return Remove(typeof(T).Name);
    }
    public bool Switch(IState state)
    {
        //如果当前状态已经是切换的目标状态,则无须切换,返回值为false
        if (CurrentState==state) return false;
        //如果当前状态不为空,则执行状态退出事件
        CurrentState?.OnExit();
        //判断切换的目标状态是否存在于列表中
        if (!states.Contains(state)) return false;
        //更新当前状态
```

```csharp
            CurrentState=state;
            //更新后,如果当前状态不为空,则执行状态进入事件
            CurrentState?.OnEnter();
            return true;
        }
        public bool Switch(string stateName)
        {
            //根据状态名称在列表中查询
            var targetState=states.Find(m =>m.Name==stateName);
            return Switch(targetState);
        }
        public bool Switch<T>() where T : IState
        {
            return Switch(typeof(T).Name);
        }
        public void Switch2Next()
        {
            if (states.Count !=0)
            {
                //如果当前状态不为空,则根据当前状态找到下一种状态
                if (CurrentState !=null)
                {
                    int index=states.IndexOf(CurrentState);
                    index=index +1<states.Count ? index +1 : 0;
                    IState targetState=states[index];
                    //首先执行当前状态的退出事件,再更新到下一状态
                    CurrentState.OnExit();
                    CurrentState=targetState;
                }
                //如果当前状态为空,则直接进入列表中的第1种状态
                else CurrentState=states[0];
                //执行状态进入事件
                CurrentState.OnEnter();
            }
        }
        public void Switch2Last()
        {
            if (states.Count !=0)
            {
                //如果当前状态不为空,则根据当前状态找到上一种状态
                if (CurrentState !=null)
```

```csharp
            {
                int index=states.IndexOf(CurrentState);
                index=index -1 >=0 ? index -1 : states.Count -1;
                IState targetState=states[index];
                //首先执行当前状态的退出事件,再更新到上一状态
                CurrentState.OnExit();
                CurrentState=targetState;
            }
            //如果当前状态为空,则直接进入列表中的最后一种状态
            else CurrentState=states[states.Count -1];
            //执行状态进入事件
            CurrentState.OnEnter();
        }
    }
    public void Switch2Null()
    {
        if (CurrentState !=null)
        {
            CurrentState.OnExit();
            CurrentState=null;
        }
    }
    public T GetState<T>(string stateName) where T : IState
    {
        return (T)states.Find(m =>m.Name==stateName);
    }
    public T GetState<T>() where T : IState
    {
        return (T)states.Find(m =>m.Name==typeof(T).Name);
    }
    public void OnUpdate()
    {
        //若当前状态不为空,则执行状态停留事件
        CurrentState?.OnStay();
        //检测所有状态切换条件
        for (int i=0; i<conditions.Count; i++)
        {
            var condition=conditions[i];
            //条件满足
            if (condition.predicate.Invoke())
            {
```

```csharp
            //如果源状态名称为空,则表示从任意状态切换至目标状态
            if (string.IsNullOrEmpty(condition.sourceStateName))
            {
                Switch(condition.targetStateName);
            }
            //如果源状态名称不为空,则表示从指定状态切换至目标状态
            else
            {
                //首先判断当前的状态是否为指定的状态
                if (CurrentState.Name==condition.sourceStateName)
                {
                    Switch(condition.targetStateName);
                }
            }
        }
    }
}
public void OnDestroy()
{
    //执行状态机内所有状态的状态终止事件
    for (int i=0; i<states.Count; i++)
    {
        states[i].OnTermination();
    }
}
public StateMachine SwitchWhen(Func<bool>predicate,
    string targetStateName)
{
    conditions.Add(new StateSwitchCondition(
        predicate, null, targetStateName));
    return this;
}
public StateMachine SwitchWhen(Func<bool>predicate,
    string sourceStateName, string targetStateName)
{
    conditions.Add(new StateSwitchCondition(
        predicate, sourceStateName, targetStateName));
    return this;
}
public StateBuilder<T>Build<T>(
    string stateName=null) where T : State, new()
```

```
    {
        Type type=typeof(T);
        T t=(T)Activator.CreateInstance(type);
        t.Name=string.IsNullOrEmpty(stateName) ? type.Name : stateName;
        if (states.Find(m =>m.Name==t.Name)==null)
            states.Add(t);
        return new StateBuilder<T>(t, this);
    }
}
```

其中,Build()方法用于使用状态构建器构建状态,状态构建器中提供了设置各状态事件和状态切换条件的接口,代码如下:

```
//第7章/StateBuilder.cs

using System;

public class StateBuilder<T>where T : State, new()
{
    private readonly T state;
    private readonly StateMachine stateMachine;

    public StateBuilder(T state, StateMachine stateMachine)
    {
        this.state=state;
        this.stateMachine=stateMachine;
    }
    public StateBuilder<T>OnInitialization(Action<T>onInitialization)
    {
        state.onInitialization=() =>onInitialization(state);
        return this;
    }
    public StateBuilder<T>OnEnter(Action<T>onEnter)
    {
        state.onEnter=() =>onEnter(state);
        return this;
    }
    public StateBuilder<T>OnStay(Action<T>onStay)
    {
        state.onStay=() =>onStay(state);
        return this;
    }
```

```csharp
    public StateBuilder<T>OnExit(Action<T>onExit)
    {
        state.onExit=() =>onExit(state);
        return this;
    }
    public StateBuilder<T>OnTermination(Action<T>onTermination)
    {
        state.onTermination=() =>onTermination(state);
        return this;
    }
    public StateBuilder<T>SwitchWhen(Func<bool>predicate,
        string targetStateName)
    {
        state.SwitchWhen(predicate, targetStateName);
        return this;
    }
    public StateMachine Complete()
    {
        state.OnInitialization();
        return stateMachine;
    }
}
```

7.3.2 敌方战斗单位 AI

创建一个新的状态机,通过状态构建器构建第 1 种状态,即巡逻状态。通过 Transform 类型数组指定巡逻的点位,通过 Navigation 实现寻路功能。

当状态进入时,将寻路的目的地设为第 1 个巡逻点位。当状态停留时,判断是否到达了巡逻目的地,到达巡逻目的地后休息指定时长,继续向下一个巡逻点位寻路。

为巡逻状态设置状态切换条件,当与 Player 的距离小于警戒范围时,切换至追击状态,代码如下:

```csharp
//第 7 章/EnemyUnitSample.cs

[SerializeField] private Transform player;
[SerializeField] private NavMeshAgent agent;
[SerializeField] private Animator animator;
//巡逻点集合
[SerializeField] private Transform[] patrolPoints;
//到达一个巡逻点后的休息时长
[SerializeField] private float restTime=3f;
```

```csharp
//警戒范围
[SerializeField] private float guardingRange=3f;
//当前巡逻点的索引值
private int index;
//休息计时器
private float restTimer;
//状态机
private StateMachine machine;

private void Start()
{
    machine=new StateMachine()
        .Build<State>("巡逻状态")
            .OnEnter(s =>
            {
                //进入巡逻状态,设置第1个巡逻点并开始巡逻
                agent.isStopped=false;
                agent.stoppingDistance=0f;
                agent.speed=1f;
                index=0;
                agent.SetDestination(
                    patrolPoints[index].position);
                animator.SetBool("Move", true);
            })
            .OnStay(s =>
            {
                //到达指定巡逻点
                if (Vector3.Distance(transform.position,
                    patrolPoints[index].position)<=.1f)
                {
                    animator.SetBool("Move", false);
                    restTimer+=Time.deltaTime;
                    //休息指定时间,随后找到下一个巡逻点进行巡逻
                    if (restTimer >= restTime)
                    {
                        restTimer=0f;
                        index++;
                        index %=patrolPoints.Length;
                        agent.SetDestination(
                            patrolPoints[index].position);
                        animator.SetBool("Move", true);
```

```
            }
        }
    })
    .OnExit(s =>
    {
        agent.isStopped=true;
        animator.SetBool("Move", false);
    })
    //假设警戒范围为3m,当与Player的距离小于3时进入追击状态
    .SwitchWhen(() =>Vector3.Distance(
        transform.position, player.position)<=guardingRange,
        "追击状态")
    .Complete();
}
```

当构建追击状态时,在状态停留事件中判断与Player的距离是否小于攻击范围时,切换至攻击状态,否则持续地将寻路目的地设置为Player的位置。

为追击状态设置状态切换条件,当Player离开警戒范围后,不再继续追击,回到巡逻状态,代码如下:

```
//第7章/EnemyUnitSample.cs

//攻击范围
[SerializeField] private float attackRange=1.2f;
private void Start()
{
    //...
    machine
        .Build<State>("追击状态")
        .OnEnter(s =>
        {
            agent.isStopped=false;
            agent.stoppingDistance=0.8f;
            //追击时的移动速度略高于巡逻时的移动速度
            agent.speed=1.5f;
            animator.SetBool("Move", true);
        })
        .OnStay(s =>
        {
            //追击过程是向Player位置寻路的过程,其目的是到Player前进行攻击
            //因此未到达Player位置时不停地进行寻路,到达后切换至攻击状态
```

```
            if (Vector3.Distance(
                    transform.position, player.position) >attackRange)
                agent.SetDestination(player.position);
            else
                s.machine.Switch("攻击状态");
        })
        .OnExit(s =>
        {
            animator.SetBool("Move", false);
        })
        //当Player离开警戒范围后回到巡逻状态
        .SwitchWhen(() =>Vector3.Distance(
            transform.position, player.position) >guardingRange,
            "巡逻状态")
    .Complete();
}
```

当构建攻击状态时，在状态停留事件中实现攻击逻辑，当 Player 离开攻击范围后，切换至追击状态。

所有状态构建完成后，调用状态机的接口进入第 1 种状态，并且在 Update( )方法中刷新状态机，代码如下：

```
//第 7 章/EnemyUnitSample.cs

//攻击计时器
private float attackTimer;
private void Start()
{
    //...
    machine
        .Build<State>("攻击状态")
            .OnEnter(s =>
            {
                agent.isStopped=true;
            })
            .OnStay(s =>
            {
                //朝向 Player
                transform.rotation=Quaternion.LookRotation(
                    player.position -transform.position);
                if (attackTimer >= 0f)
```

```
                    attackTimer -=Time.deltaTime;
                else
                {
                    if (attackTimer<=0f)
                    {
                        Debug.Log("攻击"); //TODO:攻击
                        attackTimer=2f;
                    }
                }
            })
            .SwitchWhen(() =>Vector3.Distance(
                transform.position, player.position) >attackRange,
                "追击状态")
        .Complete();
    //进入第1种状态
    machine.Switch2Next();
}
private void Update()
{
    machine.OnUpdate();
}
```

7.4 汽车驱动

本节将介绍如何在 Unity 中实现汽车的驱动功能，包括 Wheel Collider 组件的应用、车辆的转向控制、ABS 和 ASR 系统的模拟、尾气特效的制作、汽车引擎音效的播放及汽车撞击变形效果的实现等。

7.4.1 车轮碰撞器

Wheel Collider 是一种特殊的碰撞器组件，用于实现车辆驱动，它不仅可以模拟汽车的前进、后退、刹车、转向，还可以模拟汽车的悬挂系统、轮胎摩擦等物理特性，如图 7-19 所示。

虽然 Wheel Collider 组件表示车轮碰撞器，但是该组件通常不会挂载于车轮模型对象，车轮模型仅用于行驶和转向动画。在层级结构中，车轮碰撞器与车轮模型彼此独立，如图 7-20 所示，Mesh 物体的子物体是汽车 4 个车轮的模型对象，Collider 物体的子物体是 4 个车轮对应的车轮碰撞器。

Wheel Collider 类中的 GetWorldPose() 方法用于获取车轮的坐标与旋转值，将获取的坐标与旋转值赋值给车轮模型对象对应的 Transform 即可实现车轮动画，代码如下：

```csharp
//根据车轮碰撞器应用车轮的坐标及旋转
private void ApplyWheelPose(Transform w, WheelCollider wc)
{
    wc.GetWorldPose(out Vector3 pos, out Quaternion rot);
    w.position=pos;
    w.rotation=rot;
}
```

图 7-19　Wheel Collider 组件

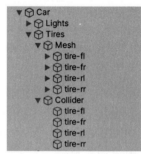

图 7-20　车轮碰撞器与车轮模型彼此独立

Wheel Collider 组件的核心属性见表 7-1。

表 7-1　Wheel Collider 组件的核心属性

属　　性	详　　解
Mass	车轮质量，单位为 kg
Radius	车轮半径
Wheel Damping Rate	车轮阻尼率
Suspension Distance	车轮悬挂最大延伸距离（向下延伸）
Force App Point Distance	车轮受力点，该值应是距车轮底部静止位置的距离，当值为 0 时表示受力点位于静止的车轮底部，较好的车辆会使受力点略低于车辆质心

续表

属　性	详　解
Center	车轮中心点
Suspension Spring	车轮悬挂相关属性
Spring	弹簧，值越大悬挂到达目标位置（Target Position）的速度越快
Damper	阻尼，值越大悬挂弹簧移动越慢
Target Position	悬挂的静止距离，0 表示完全压缩的悬挂，1 表示完全展开的悬挂
Forward/Sideways Friction	车轮前向/侧向摩擦力的相关属性
Extremum Slip/Value	摩擦曲线的极值点
Asymptote Slip/Value	摩擦曲线的渐近点
Stiffness	改变摩擦力的刚度，设为 0 表示完全禁用车轮的所有摩擦力

轮胎摩擦通过摩擦曲线表示，如图 7-21 所示，曲线以轮胎打滑的度量值作为输入，并以力作为输出，在低打滑条件下，轮胎会施加较大的力，当打滑变高时，随着轮胎开始滑动或旋转，力会减小。

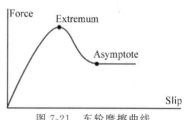

图 7-21　车轮摩擦曲线

通过 Wheel Collider 类中的 motorTorque 和 brakeTorque 属性可以实现车辆的驱动和制动，motorTorque 表示车轮的驱动力矩，brakeTorque 表示车轮的刹车力矩，单位均为 N·m。除此之外，steerAngle 属性表示车轮的转向角度，用于实现车轮的转向，以度为单位。

值得注意的是，车辆需要添加刚体组件，并且车辆的碰撞器不能包裹住 Wheel Collider，另外，为了防止车辆发生侧翻，通常会稍微降低其刚体的质心，如图 7-22 所示。

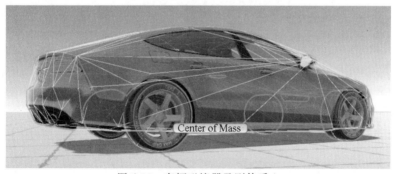

图 7-22　车辆碰撞器及刚体质心

7.4.2　驱动类型

车辆驱动分为前驱、后驱和四驱 3 种类型，当获取垂直方向上的输入时，根据输入值为

相应的车轮施加驱动扭矩,假设通过空格键触发车辆制动,那么当获取空格键输入时,为车轮设置制动扭矩,代码如下:

```csharp
//第 7 章/CarController.cs

using UnityEngine;

public class CarController : MonoBehaviour
{
    private Rigidbody rb;
    [SerializeField] private Transform com;

    [Header("Wheel")]
    //车轮
    [SerializeField] private Transform w_fl;
    [SerializeField] private Transform w_fr;
    [SerializeField] private Transform w_rl;
    [SerializeField] private Transform w_rr;

    //车轮碰撞器
    [SerializeField] private WheelCollider wc_fl;
    [SerializeField] private WheelCollider wc_fr;
    [SerializeField] private WheelCollider wc_rl;
    [SerializeField] private WheelCollider wc_rr;

    [Header("Engine")]
    [SerializeField]
    private DriverType driverType=DriverType.All;
    //驱动类型(前驱、后驱、四驱)
    public enum DriverType { Front, Rear, All };

    [SerializeField]
    private float motorTorqueFactor=2000f;
    [SerializeField]
    private float brakeTorqueFactor=1000f;
    private float currentMotorTorque;
    private float currentBrakeTorque;

    private void Start()
    {
        rb=GetComponent<Rigidbody>();
```

```csharp
        rb.centerOfMass=com.localPosition;
}

private void FixedUpdate()
{
    currentBrakeTorque=Input.GetKey(KeyCode.Space)
        ?brakeTorqueFactor : 0f;
    ApplyBrakeTorque(wc_fl, currentBrakeTorque);
    ApplyBrakeTorque(wc_fr, currentBrakeTorque);
    ApplyBrakeTorque(wc_rl, currentBrakeTorque);
    ApplyBrakeTorque(wc_rr, currentBrakeTorque);

    currentMotorTorque=currentBrakeTorque!=0f
        ? 0f
        : Input.GetAxis("Vertical") * motorTorqueFactor;
    switch(driverType)
    {
        case DriverType.Front:
            ApplyMotorTorque(wc_fl, currentMotorTorque);
            ApplyMotorTorque(wc_fr, currentMotorTorque);
            break;
        case DriverType.Rear:
            ApplyMotorTorque(wc_rl, currentMotorTorque);
            ApplyMotorTorque(wc_rr, currentMotorTorque);
            break;
        case DriverType.All:
            ApplyMotorTorque(wc_fl, currentMotorTorque * .5f);
            ApplyMotorTorque(wc_fr, currentMotorTorque * .5f);
            ApplyMotorTorque(wc_rl, currentMotorTorque * .5f);
            ApplyMotorTorque(wc_rr, currentMotorTorque * .5f);
            break;
    }

    ApplyWheelPose(w_fl, wc_fl);
    ApplyWheelPose(w_fr, wc_fr);
    ApplyWheelPose(w_rl, wc_rl);
    ApplyWheelPose(w_rr, wc_rr);
}
//应用驱动扭矩
private void ApplyMotorTorque(WheelCollider wc, float motorTorque)
{
```

```
        wc.motorTorque=motorTorque;
    }
    //应用制动扭矩
    private void ApplyBrakeTorque(WheelCollider wc, float brakeTorque)
    {
        wc.brakeTorque=brakeTorque;
    }
    //根据车轮碰撞器应用车轮的坐标及旋转
    private void ApplyWheelPose(Transform w, WheelCollider wc)
    {
        wc.GetWorldPose(out Vector3 pos, out Quaternion rot);
        w.position=pos;
        w.rotation=rot;
    }
}
```

7.4.3 车辆转向

车辆的转向通过前车轮控制,当获取水平方向上的输入时,根据输入值为两个前车轮设置转向角度,代码如下

```
//第7章/CarController.cs

[SerializeField]
private float steeringAngleLimit=45f;
private float currentSteeringAngle;

private void FixedUpdate()
{
    //...
    currentSteeringAngle=Input.GetAxis("Horizontal") * steeringAngleLimit;
    wc_fl.steerAngle=currentSteeringAngle;
    wc_fr.steerAngle=currentSteeringAngle;
    //...
}
```

效果如图 7-23 所示。

7.4.4 行驶速度

Wheel Collider 类中的 rpm 属性表示轮轴的转速,可以用于计算车速,以每分钟转数为单位。车轮的周长可以通过车轮的半径求得,将车轮的周长乘以 rpm 得到每分钟行驶的距离,车速以 km/h 为单位,因此最终乘以 60 并除以 1000 便可以得到车速,代码如下:

图 7-23 车轮转向

```
//第7章/CarController.cs

private float km; //车速 km/h

private void FixedUpdate()
{
    //...
    float circumference=wc_rl.radius * 2 * Mathf.PI;
    float rpm=(Mathf.Abs(wc_rl.rpm) +Mathf.Abs(wc_rr.rpm)) * .5f;
    km=Mathf.Round(circumference * rpm * 60 * .001f);
    //...
}
```

7.4.5 ABS 与 ASR

制动防抱死系统(Antilock Brake System,ABS)当紧踩刹车过程中遇到打滑情况时,轮胎的滚动摩擦变为滑动摩擦,车轮直接抱死,不仅刹不住车,方向也会失控,而 ABS 系统会在刹车时自动执行点刹车,防止车辆在紧急制动时车轮抱死。

驱动防滑系统(Anti-Slip Regulation,ASR)没有搭载 ASR 的车辆在光滑路面上加速时,驱动轮容易打滑,后驱车辆可能出现甩尾,前驱车辆则容易方向失控,导致车辆侧向滑移,而 ASR 可以使驱动轮的滑转率在最佳范围内,从而可避免在加速时驱动轮打滑,并保证车辆转向能力,同时可以减少轮胎磨损。

轮胎的打滑情况可以通过 Wheel Collider 类中的 GetGroundHit()方法获取,该方法用于获取车轮的地面碰撞数据,这些数据存储在 Wheel Hit 类型的结构体中,Wheel Hit 中包含的核心属性见表 7-2。

表 7-2 Wheel Hit 中包含的核心属性

属 性	详 解
collider	车轮所撞击的其他碰撞器
force	施加到接触面的力的大小

属　性	详　解
forwardDir	车轮的前向方向
forwardSlip	前向方向的轮胎打滑，加速打滑为负数，制动打滑为正数
normal	接触点处的法线
point	车轮与地面之间的接触点
sidewaysDir	车轮的侧向方向
sidewaysSlip	侧向方向的轮胎打滑

在应用制动扭矩或驱动扭矩时获取车轮的地面碰撞数据，如果轮胎打滑大于一定阈值，则将制动扭矩或驱动扭矩设为0，代码如下：

```
//第 7 章/CarController.cs

//应用驱动扭矩
private void ApplyMotorTorque(WheelCollider wc, float motorTorque)
{
    //wc.motorTorque=motorTorque;
    wc.motorTorque=ASR
        && wc.GetGroundHit(out WheelHit hit)
        && hit.forwardSlip > .35f
            ? 0f
            : motorTorque;
}

//应用制动扭矩
private void ApplyBrakeTorque(WheelCollider wc, float brakeTorque)
{
    //wc.brakeTorque=brakeTorque;
    wc.brakeTorque=ABS
        && wc.GetGroundHit(out WheelHit hit)
        && hit.forwardSlip > .35f
            ? 0f
            : brakeTorque;
}
```

当然，此处仅仅通过控制驱动力和制动力来模拟 ASR 和 ABS 系统，在实际应用中，这些系统通常涉及复杂的硬件和软件集成，并需要符合严格的安全和性能标准。

7.4.6　尾气排放

尾气排放效果通过粒子系统实现，实现该粒子系统首先要准备一张烟雾效果的贴图，如

图 7-24 所示,然后创建一个使用 Particles/Alpha Blended 类型 Shader 的材质球,为其设置烟雾贴图,最终将该材质球应用于粒子系统。

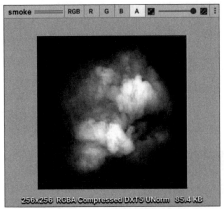

图 7-24　烟雾贴图

粒子的发射频率可以在脚本中动态调整,例如在车辆起步过程中将发射频率设置得较大,当车速大于一定阈值后停止发射,代码如下:

```csharp
//第 7 章/CarController.cs

[SerializeField]
private ParticleSystem[] smokes;

private void FixedUpdate()
{
    //...
    for (int i=0; i<smokes.Length; i++)
    {
        ParticleSystem smoke=smokes[i];
        if (km<=20f)
        {
            var emission=smoke.emission;
            if (!emission.enabled)
                emission.enabled=true;
            emission.rateOverTime=Input.GetAxis("Vertical") >.1f
                ? 20 : 5;
        }
        else
        {
            if (smoke.emission.enabled)
            {
```

```
            var emission=smoke.emission;
            emission.enabled=false;
        }
    }
}
```

将粒子系统的 Start Size、Start Rotation 等参数调节到合适的大小，如图 7-25 所示，为了实现气体的消散效果，在 Color over Lifetime 中将生命周期结束时的颜色透明度设为 0，如图 7-26 所示。

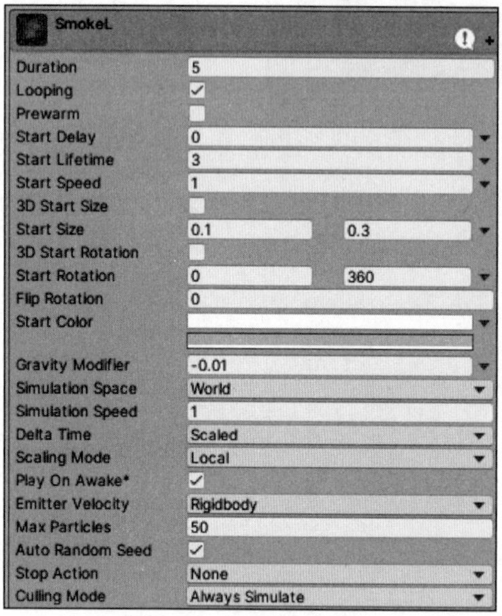

图 7-25　烟雾粒子特效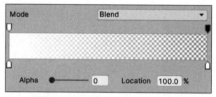

图 7-26　Color over Lifetime

播放粒子特效，效果如图 7-27 所示。

图 7-27　尾气排放效果

7.4.7 车辆音效

准备相关音频剪辑,例如车辆发动机运行所需的音频、车辆制动所需的音频等,创建空物体,添加 Audio Source 组件,以便播放这些音频剪辑,如图 7-28 所示。

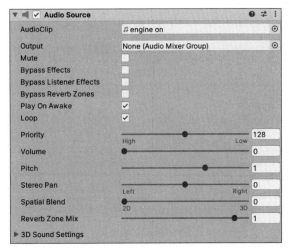

图 7-28 Audio Source

在车辆行驶过程中,根据车速设置车辆发动机声音的 Volume 和 Pitch 大小,而对于车辆制动声音的大小,则需要根据用户当前是否按下空格键设置,示例代码如下:

```
//第 7 章/CarController.cs

[SerializeField] private AudioSource engineOnSound;
[SerializeField] private AudioSource brakeSound;

private void Sounds()
{
    engineOnSound.volume=Mathf.Lerp(engineOnSound.volume,
        Mathf.Lerp(.2f, 1f, km / 255f),
        Time.deltaTime * 15f);
    engineOnSound.pitch=Mathf.Lerp(engineOnSound.pitch,
        Mathf.Lerp(.5f, 1.5f, km / 255f),
        Time.deltaTime * 15f);

    brakeSound.volume=Mathf.Lerp(0f, 1f,
        Input.GetKey(KeyCode.Space) ? km / 255f : 0f);
}
```

7.4.8 撞击变形

两个游戏物体发生碰撞有两个必要条件,一是发生碰撞的两个游戏物体包含碰撞器组件,二是其中一个游戏物体包含刚体组件。当两个游戏物体发生碰撞时,MonoBehaviour 中的相关回调函数会被调用,详解见表 7-3。

表 7-3 MonoBehaviour 中碰撞相关的回调函数

函 数	详 解
OnCollisionEnter(Collision collision)	当前碰撞器开始接触另一个碰撞器时被调用
OnCollisionStay(Collision collision)	当前碰撞器持续接触另一个碰撞器时被调用
OnCollisionExit(Collision collision)	当前碰撞器停止接触另一个碰撞器时被调用

物体发生碰撞时的碰撞数据存储于 collision 对象中,其中包含的核心属性见表 7-4。

表 7-4 Collision 类中的核心属性

属 性	详 解
articulationBody	与当前碰撞器发生碰撞的物体上的 Articulation Body 组件
body	与当前碰撞器发生碰撞的物体上的 Rigidbody 或 Articulation Body 组件
collider	与当前碰撞器发生碰撞的碰撞器
contactCount	碰撞接触点的数量
contacts	碰撞接触点数组,碰撞接触点包含位置和法线数据
gameObject	与当前碰撞器发生碰撞的物体
impulse	碰撞冲量
relativeVelocity	两个碰撞对象的相对线性速度
rigidbody	与当前碰撞器发生碰撞的物体上的 Rigidbody 组件
transform	与当前碰撞器发生碰撞的物体的 Transform 组件

在 OnCollisionEnter() 方法中获取碰撞数据,如果车辆与碰撞对象的相对线性速度大于一定阈值,则可以判定车辆发生了撞击,通过调整车辆模型的网格顶点,使车辆外观变形。网格顶点的位移量与碰撞接触点的距离相关,示例代码如下:

```
//第 7 章/CarController.cs

[SerializeField] private float deformationThreshold=5f;
[SerializeField] private float deformationRadius=.5f;

private void OnCollisionEnter(Collision collision)
{
    //发生撞击
    if (collision.relativeVelocity.magnitude >=deformationThreshold)
    {
```

```csharp
Mesh mesh=GetComponent<MeshFilter>().mesh;
Vector3[] vertices=mesh.vertices;
for (int i=0; i<collision.contactCount; i++)
{
    ContactPoint contactPoint=collision.GetContact(i);
    Vector3 world2Local=transform
        .InverseTransformPoint(contactPoint.point); //转局部坐标
    for (int j=0; j<vertices.Length; j++)
    {
        float magnitude=(world2Local -vertices[j]).magnitude;
        if (magnitude<deformationRadius)
        {
            float delta=(deformationRadius -magnitude)
                / deformationRadius * .01f;
            vertices[j]+=transform.InverseTransformDirection(
                collision.relativeVelocity) * delta;
        }
    }
}
//更新网格顶点
mesh.vertices=vertices;
mesh.RecalculateNormals();
mesh.RecalculateBounds();
//重启碰撞
var collider=GetComponent<Collider>();
collider.enabled=false;
collider.enabled=true;
```

运行程序,驾驶车辆撞击其他物体,结果如图 7-29 所示。

图 7-29　撞击变形

图 书 推 荐

书 名	作 者
仓颉语言实战（微课视频版）	张磊
仓颉语言核心编程——入门、进阶与实战	徐礼文
仓颉语言程序设计	董昱
仓颉程序设计语言	刘安战
仓颉语言元编程	张磊
仓颉语言极速入门——UI 全场景实战	张云波
HarmonyOS 移动应用开发（ArkTS 版）	刘安战、余雨萍、陈争艳 等
公有云安全实践（AWS 版·微课视频版）	陈涛、陈庭暄
虚拟化 KVM 极速入门	陈涛
虚拟化 KVM 进阶实践	陈涛
移动 GIS 开发与应用——基于 ArcGIS Maps SDK for Kotlin	董昱
Vue＋Spring Boot 前后端分离开发实战（第 2 版·微课视频版）	贾志杰
前端工程化——体系架构与基础建设（微课视频版）	李恒谦
TypeScript 框架开发实践（微课视频版）	曾振中
精讲 MySQL 复杂查询	张方兴
Kubernetes API Server 源码分析与扩展开发（微课视频版）	张海龙
编译器之旅——打造自己的编程语言（微课视频版）	于东亮
全栈接口自动化测试实践	胡胜强、单镜石、李睿
Spring Boot＋Vue.js＋uni-app 全栈开发	夏运虎、姚晓峰
Selenium 3 自动化测试——从 Python 基础到框架封装实战（微课视频版）	栗任龙
Unity 编辑器开发与拓展	张寿昆
跟我一起学 uni-app——从零基础到项目上线（微课视频版）	陈斯佳
Python Streamlit 从入门到实战——快速构建机器学习和数据科学 Web 应用（微课视频版）	王鑫
Java 项目实战——深入理解大型互联网企业通用技术（基础篇）	廖志伟
Java 项目实战——深入理解大型互联网企业通用技术（进阶篇）	廖志伟
深度探索 Vue.js——原理剖析与实战应用	张云鹏
前端三剑客——HTML5＋CSS3＋JavaScript 从入门到实战	贾志杰
剑指大前端全栈工程师	贾志杰、史广、赵东彦
JavaScript 修炼之路	张云鹏、戚爱斌
Flink 原理深入与编程实战——Scala＋Java（微课视频版）	辛立伟
Spark 原理深入与编程实战（微课视频版）	辛立伟、张帆、张会娟
PySpark 原理深入与编程实战（微课视频版）	辛立伟、辛雨桐
HarmonyOS 原子化服务卡片原理与实战	李洋
鸿蒙应用程序开发	董昱
HarmonyOS App 开发从 0 到 1	张诏添、李凯杰
Android Runtime 源码解析	史宁宁
恶意代码逆向分析基础详解	刘晓阳
网络攻防中的匿名链路设计与实现	杨昌家
深度探索 Go 语言——对象模型与 runtime 的原理、特性及应用	封幼林
深入理解 Go 语言	刘丹冰
Spring Boot 3.0 开发实战	李西明、陈立为

续表

书　名	作　者
全解深度学习——九大核心算法	于浩文
HuggingFace 自然语言处理详解——基于 BERT 中文模型的任务实战	李福林
动手学推荐系统——基于 PyTorch 的算法实现（微课视频版）	於方仁
深度学习——从零基础快速入门到项目实践	文青山
LangChain 与新时代生产力——AI 应用开发之路	陆梦阳、朱剑、孙罗庚、韩中俊
图像识别——深度学习模型理论与实战	于浩文
编程改变生活——用 PySide6/PyQt6 创建 GUI 程序（基础篇·微课视频版）	邢世通
编程改变生活——用 PySide6/PyQt6 创建 GUI 程序（进阶篇·微课视频版）	邢世通
编程改变生活——用 Python 提升你的能力（基础篇·微课视频版）	邢世通
编程改变生活——用 Python 提升你的能力（进阶篇·微课视频版）	邢世通
Python 量化交易实战——使用 vn.py 构建交易系统	欧阳鹏程
Python 从入门到全栈开发	钱超
Python 全栈开发——基础入门	夏正东
Python 全栈开发——高阶编程	夏正东
Python 全栈开发——数据分析	夏正东
Python 编程与科学计算（微课视频版）	李志远、黄化人、姚明菊 等
Python 数据分析实战——从 Excel 轻松入门 Pandas	曾贤志
Python 概率统计	李爽
Python 数据分析从 0 到 1	邓立文、俞心宇、牛瑶
Python 游戏编程项目开发实战	李志远
Java 多线程并发体系实战（微课视频版）	刘宁萌
从数据科学看懂数字化转型——数据如何改变世界	刘通
Dart 语言实战——基于 Flutter 框架的程序开发（第 2 版）	亢少军
Dart 语言实战——基于 Angular 框架的 Web 开发	刘仕文
FFmpeg 入门详解——音视频原理及应用	梅会东
FFmpeg 入门详解——SDK 二次开发与直播美颜原理及应用	梅会东
FFmpeg 入门详解——流媒体直播原理及应用	梅会东
FFmpeg 入门详解——命令行与音视频特效原理及应用	梅会东
FFmpeg 入门详解——音视频流媒体播放器原理及应用	梅会东
FFmpeg 入门详解——视频监控与 ONVIF＋GB28181 原理及应用	梅会东
Python 玩转数学问题——轻松学习 NumPy、SciPy 和 Matplotlib	张骞
Pandas 通关实战	黄福星
深入浅出 Power Query M 语言	黄福星
深入浅出 DAX——Excel Power Pivot 和 Power BI 高效数据分析	黄福星
从 Excel 到 Python 数据分析：Pandas、xlwings、openpyxl、Matplotlib 的交互与应用	黄福星
云原生开发实践	高尚衡
云计算管理配置与实战	杨昌家
HarmonyOS 从入门到精通 40 例	戈帅
OpenHarmony 轻量系统从入门到精通 50 例	戈帅
AR Foundation 增强现实开发实战（ARKit 版）	汪祥春
AR Foundation 增强现实开发实战（ARCore 版）	汪祥春